Richard Hughes

The knowledge of the physician

A course of lectures delivered at the Boston university school of medicine, May,

1884

Richard Hughes

The knowledge of the physician
A course of lectures delivered at the Boston university school of medicine, May, 1884

ISBN/EAN: 9783337257910

Printed in Europe, USA, Canada, Australia, Japan

Cover: Foto ©berggeist007 / pixelio.de

More available books at **www.hansebooks.com**

THE

KNOWLEDGE OF THE PHYSICIAN.

A Course of Lectures

DELIVERED AT THE BOSTON UNIVERSITY SCHOOL OF
MEDICINE, MAY, 1884.

BY

RICHARD HUGHES, M.D.

———————

BOSTON:

OTIS CLAPP AND SON.

1884.

ELECTROTYPED AND PRINTED
BY RAND, AVERY, AND COMPANY,
BOSTON, MASS.

TO

I. T. TALBOT, M.D.,

DEAN

OF THE BOSTON UNIVERSITY SCHOOL OF MEDICINE,

𝔗𝔥𝔦𝔰 𝔙𝔬𝔩𝔲𝔪𝔢 𝔦𝔰 𝔇𝔢𝔡𝔦𝔠𝔞𝔱𝔢𝔡

BY HIS FRIEND AND SINCERE APPRECIATOR,

THE AUTHOR.

CONTENTS.

v

LECTURE III.

THE KNOWLEDGE OF DISEASE.

LECTURE IV.

THE KNOWLEDGE OF MEDICINES.

LECTURE V.

PYREXIA AND THE ANTIPYRETICS.

THE

KNOWLEDGE OF THE PHYSICIAN.

I.

THE KNOWLEDGE OF LIFE.

Ladies[1] and Gentlemen, — It is, believe me, with
keen pleasure that I stand here to-day. American
homœopathy has always enlisted my warm sympa-
thies; but it has been with especial interest that I
have watched the rise and progress of the Boston
University School of Medicine, now in the second
decade of its existence. That such a school should
exist, without distinctive name, pledged to the
method of Hahnemann only by the convictions of
its teachers and the preferences of its pupils, — this
makes actual my ideal of what should be. I would
it were possible in my own country: I am very glad
that it has become so in yours. While, therefore, I
feel it an honour and pleasure to be invited to address
American students, I am particularly gratified that
Boston should be the place in which I am to speak.

[1] The students of Boston University are of both sexes.

1

In requesting me to deliver this course of lectures, my friends here have expressed certain wishes as to the topics of which they should treat. Their main subject-matter (it is allowed) should be that to which I have given the best work of my life, and in the sphere of which I have made my poor contribution to our knowledge, viz. Materia Medica and Thera- peutics. But it has been thought well that discussion on this head should be preceded by some consider- ations of a more general nature, bearing on Medicine as a whole. In deciding what these should be, I have been determined mainly by the nature of my audience. You are learners as yet, rather than practitioners : what you should *know* concerns you at present more that what you should *do*. At the same time, it would be a superfluous task to seek to augment your actual knowledge of the topics of medical education. To go again over ground which you have trodden under the leadership of such men as adorn this School would be labour in vain : if you have not learned — or are not learning — from them the physiology, the pathology, the pharmacology they teach, you would not learn it from another. But it has seemed to me that I might to your advantage draw you back a lit- tle from your fields of work, and ask you to take with me a survey of the country at large, — to enquire what is the knowledge you are seeking, both in itself, and in relation to the several departments to which it belongs and the action in which it ends. You are to practise the art of healing : you are acquiring the sciences accessory thereto, that you may fulfil your calling aright. What should you know of all of these? what of each?

Every science — that is, branch of knowledge, *scientia* from *scire* — consists of three parts, phenomena, laws, and causes. You may perhaps be thinking that phenomena are not peculiar to science, that they are matters of common observation. And so they are; but common observation just differs from skilled, in that it very imperfectly takes them in. Night after night the host of stars circle around the pole, and the moon and planets wander about among them. All eyes perceive this; but how few minds take in the details of their order and march! how few, for instance, when poets like Dante and Tennyson indicate the time of year by the positions then occupied by the constellations, can recognise the description and seize the indication! He who has studied these and such-like phenomena has taken the first step towards acquaintance with the science of astronomy. And he has done more. He has mastered a series of facts which henceforth are part of the furniture of his mind, and can be turned to divers uses. Their perception necessarily affords pleasure, which all men can feel for themselves, and some can, by reproduction of its sources, communicate to others. The poet can so describe the starry heavens, the painter so depict them, that quite a new sense of their wonder and beauty is wakened within men. It is thus that fine art lays hold of the knowledge of phenomena. But useful art employs it no less. One of the blessings which the Prometheus of

Æschylus claims to have brought to mankind is
that he taught them

"The rising and the setting of the stars."

From observation of these, of the sun's path among
the stars, of the place of the pole and the phases
of the moon, our progenitors drew a number of
practical applications, to agriculture, to navigation,
to the ordering of life and work, which made the
lights of heaven — long before any real astronomical
science had come into being — of large service to
man, availing "for signs and for seasons, for days
and years."

I would lay the more stress on the value of knowledge of this kind, as it seems to me to have lost its
place of late, alike in education and in appreciation.
The world has been so fascinated by the discoveries
of the last few centuries that these alone occupy its
thoughts ; and to teach science means to impart
knowledge of laws and causes, with very little reference to the phenomena which those laws regulate
and for which those causes account. It is the same
tendency as that which dilates on the philosophy of
history, disregarding its facts ; as that which makes
books about great authors widely read, while few
know the original works themselves. We shall see
hereafter how disastrous such procedure has been in
the study of medicine : at present we dwell on it
only in relation to the subject of physical knowledge
in general. Whatever else you learn, inform yourselves thoroughly as to *facts*, — as to all that can be

elicited by observation and experiment. Nor take them only from teachers and books. See, hear, feel for yourselves: handle the instruments by which modern observation brings the distant near and the invisible into view, while at the same time you take in the great world of motion and sound, of colour and form, which is open every day to the unaided senses. Daily experience will warrant this counsel by the delight and profit it brings.

But while I make thus much of the knowledge of phenomena — of the exercise upon nature of the perceptive faculties, I would not undervalue that of the intellectual. Man is made to enquire, to ask "how?" and "why?" as well as "what?" The answer to the two former questions constitutes what is called inductive science — the body of inferences more or less certain, of theories with stronger or feebler warranty, which are the result of thought and imagination when in the presence of natural facts. What a crowd of splendid conceptions rises in the mind when this reference is made! The heliocentrism of Copernicus, Kepler's laws of the heavenly motions, Newton's gravitation, the nebular hypothesis of Laplace, the undulatory theory of light and the dynamic of heat, morphological type and evolutionary growth in animated nature, — these are but a few of the most striking among them. They are all thoughts rather than observations; of no necessary or even demonstrable truth; open at any time to reconsideration on adequate challenge; — but, while they stand, part of the world's richest wealth, lumi-

nous in effect and indispensable to further progress.
Because you are thinkers, and not merely perceivers,
you must enter into these great thinkings of the
giants of scientific discovery: from the vision of
things as they appear you must rise to that which
strives to see them as they are.

These principles apply to knowledge in general:
they characterise no less, of course, that knowledge
which is necessary to the physician. He is, as we
have said, the practitioner of an art; and he might
carry out his calling by pure empiricism and rule of
thumb. At first he had so to do. But, by degrees,
the life he has to guard and keep in health, the
disease he has to combat, and the drugs he has to
employ, have become the subject each of a science
special to itself. Of these sciences — of physiology,
of pathology, of pharmacology — you have to learn
the phenomena, as ascertained by observation and
experiment; and no less to possess yourselves of the
noumena, or what are believed to be such — the
laws and causes of their workings as brought into
view by intellectual vision. [I should say that by this
word "noumena" — from νοῦς, mind — philosophy
designates the reality of things as distinguished from
their appearances. This, however, is a metaphysical
conception, — one of thought rather than fact; and
I am using the word more in its etymological sense
to distinguish that which the νοῦς perceives from
that which the senses experience — both these being
(probably) realities.] Let us pass in review to-day
the phenomena and noumena of the first of these

sciences, or rather of that which is the subject of
physiology — of *life*.

The fundamental characteristic of that human body
for which, as physicians, you will have to care, is
that it is a living organism. The meaning of the
latter term is obvious. It implies that the body is
not simple, as a stone is, but complex in form and
structure, its several parts so differentiated as to
constitute distinct mechanisms capable of perform-
ing various functions, and yet all bound together in
unity of existence and (in some measure) of govern-
ment. This it is to be an organism ; but what is it
for such organism to be alive ? A musical instrument, a watch, a printing-machine, answers to the
definition just given : either is — or may be — of
complex structure, having diverse parts subserving
diverse purposes, but neither is living. What do we
mean by characterising the human organism by this
adjective ?

We mean that it belongs to a class of existences,
which includes all other animals and the whole vegetable kingdom, but excludes the world of earth and
water. In all ages mankind has recognised the dis-
tinctness of this class of beings, and has separated
them off by ascribing to them the possession of a
peculiar quality, which it has called ζωή, *vita*, life.
Plants and animals have life, they are living things, —
this is the language in which mankind expresses a
universal perception. The thought has sometimes
been confused by the use of terms implying vitality
to designate inorganic objects. The poet may indeed
pardonably speak (like Shelley) of

> " The living winds, which flow
> Like waves above the living waves below,"

or (like Virgil) of "*vivo* sedilia saxo." But when
philosophers tell us that minerals " have a life of their
own," they are employing language which is need-
lessly misleading. Minerals *exist*, having their own
laws and history ; but we want to describe that which
is peculiar to plants and animals when we say that
they *live*.

Further consideration establishes what it is to live,
wherein the *differentia* of living things resides. It is
not that they have a definite form, into which they
inevitably grow, and to which they compel all increase
of their substance ; for this property they enjoy in
common with the lifeless crystal. It is not that they
are the seat of motion. There is no rest anywhere
in nature : if to move were to live, all things would
be alive. The earth and its sister planets are revolving on their axes, and whirling round the sun ; the
sun itself and its brother luminaries are revolving on
their axes, and circling some central point in space ;
the ether is throbbing with undulations of light and
the air with those of sound ; within the earth, the
ocean is ebbing and flowing upon the shore, the rivers running to the sea, the winds above as restless
as the tides below, the " clouds which far above us
float and pause " ever resolving themselves and forming anew, heat is radiating and being absorbed, electricity is producing its attractions and repulsions, its
chemic combinations and decompositions, and — if

we are to believe its latest imaginings — keeping
every atom of matter in as restless a whirl as that
which urges the suns and their worlds. Motion is
not life; nor does it become so by being spontaneous.
We shall see hereafter that vital motions are not ini-
tiations, but responses: yet, were they otherwise,
that would not make them living, since the total motion of the universe must — so far as science can
pierce — be thus conceived of. (I mean, that you
must go outside the universe to find its Mover. In
itself it appears as self-moved.) Nor does life consist in growth and waste. We open our ears and
hear

> " The sound of streams that, swift or slow,
> Draw down Æonian hills, and sow
> The dust of continents to be."

Every wave upon the cliff, every shower of rain upon
the land, is reducing one part of the earth to enrich
another. In some places the land has risen, in others
it has sunk, within the period of man's existence;
and the whole geologic history before him is one of
continual growth and waste.

While, however, growth by accretion is thus a physical process, there is a growth which is purely vital.
A granite rock to which chalk has been brought by
rivers has no power of changing calcareous into gran-
itic substance. But the plant and the animal, diverse
as may be the structures of their several organs and
the food supplied to them, can so transform this
food within their bodies that it shall become homoge-

neous with themselves, and go to make their wood, leaf, bone and muscle, and contribute to the texture of such special organs as the brain and the eye. There is nothing like this, save in living beings : it is a purely vital process. And as their growth is not mere accretion, so the correlative waste in them is more than separation. Matter is thrown off from the tissues in lower forms than those in which it exists there : muscular fibre is eliminated as urea, nervous substance in the shape of phosphates.

Growth by assimilation is thus the first process characteristic of life. And the second is like unto it, — it is the reproduction of the kind. This is a power peculiar to plants and animals, and nothing bearing the remotest resemblance thereto exists in any but living beings. It also, therefore, is purely vital, and constitutes an element in life.

Take now the whole life-history of a plant, and we shall find it comprised in these two processes — growth by assimilation and reproduction of its kind, the means for the preservation of the individual and the perpetuation of the species. All else about it is physical, common to the world around : these are peculiar, are vital. Since, then, any property which does not belong to the vegetable creation cannot be essential to life as such, we may consider these two as making up the manifestation of that which we name vitality. Pressing our analysis farther, we find the duality resolving itself into unity ; for in the lowest forms of life generation is no distinctive process, but a simple fission or gemmation — a portion of the liv-

ing mass becoming detached, and then growing by
assimilation into the parent form. (Microscopically,
indeed, reproduction and growth are in all cases
identical, consisting essentially in the cleavage and
duplication of cells ; but I speak of them as they appear to the common eye.) Now in these primitive
regions, among the dim beginnings of things, another
fact is pressed upon our notice. It is that the essential functions of life require for their basis none of
the developments seen in the higher plants or animals. Tracing these down as they slope towards a
fellow, if not a common, root, we come at length to
creatures like the amœba. We have here neither
colour nor (permanent) form ; solidity and structure
alike are absent ; and yet no one can doubt that the
amœba lives. Retracing then our steps, we find that
as we ascend the scale of being, either along the
vegetable or the animal side of the slope, we always
carry what we may call amœboid matter with us. Behind all shapely form, within all solid envelopes, is
the structureless, transparent, colourless, semi-fluid
substance we have seen floating in the waters. We
call it *protoplasm* — not a good word, for it is rather
the first to form than the first to be formed. But,
however it be named, it must be conceived of as the
seat, and the only seat, of vitality — this property
being fully manifested when it alone is present.[1] Of
ourselves we may say, *non omnis vivo :* we are not

---

[1] " Wherever there is life, from its lowest to its highest manifestations,
there is protoplasm ; wherever there is protoplasm, there also is life " (PRO-
FESSOR ALLMAN, *Brit. Assoc. Report*, 1879).

alive in every part. The fluids of the body do not
live, any more than water does. Hair, nails, teeth,
bone, are when once formed as lifeless as similar
matter existing elsewhere ; and the same is demon-
strably, though not so obviously, true of cartilage,
fibre, cuticle, and all cell-walls and sheaths.. The soft
matter which most of these contain — the contents
of the cell, the pulp of teeth, the marrow of bone —
this lives and forms its envelope ; but the thing
formed, that which has structure, rigidity, colour, has
passed from life to death. As dead, it is continually
decaying, but as continually being replaced by its
indwelling protoplasm, which takes from the food,
reaching it as blood, the material for the purpose.

Life is thus a property belonging to matter in that
peculiar state which we denote by terming it "living ;" and such living matter, though everywhere
present in the organism, constitutes but a small part
of the bulk of the frame. The first to promulgate
this doctrine was one whom we of the school of
Hahnemann may fairly claim for our own ; for though
he never practised, and died before homœopathy became a living thing in his country, he welcomed it
and vindicated its claims to attention. I speak of
Dr. John Fletcher, of Edinburgh. His *Rudiments
of Physiology*, published between 1835 and 1837,
start from the theory first propounded by Brown in
1780, that life is the result of the action of the
natural stimuli — light, heat, oxygen and so forth —
on the "irritability" of the tissues. Their capacity
of responding thereto is vitality : their actual re-

sponse is *vita* — life. This view — espoused by all
the great naturalists and physiologists who had
followed, by Blumenbach, Cuvier, Bichat, Magendie,
Pritchard — is maintained by Fletcher, and opposed to
all theories of a "vital principle," in a series of chapters which, as Dr. Drysdale justly says, "for learning,
wit, close reasoning, and splendour of diction, have
been seldom equalled, and, I venture to say, never sur-
passed in the literature of physiology." But Fletcher
went farther, in perceiving and meeting the difficulty
that much of the organism seemed to be irresponsive
to stimuli. He did so by arguing that it is indeed
not wholly alive ; that its fluids and hard solids have
no more vitality than water or wood. The old phrase
"materia vitæ diffusa" after all expressed the truth ;
only it was a "materia" indeed, and nothing semi-
spiritual — a living matter everywhere found where
life is manifested, and constituting the substratum of
its phenomena. That such a matter there must be,
he clearly discerned and irrefragably proved. But he
unfortunately went farther still, and thought he had
found it in the ganglionic nerve-substance. This error
caused his work to be doubtfully received and soon
forgotten. There were two of his pupils, however,
who held his memory dear, and waited for the time
when his discernment would be vindicated. Both
have adorned our ranks. One was Dr. Rutherford
Russell, a man of no common power and charm, too
early lost to us. The other I have already named,
and to name him is sufficient : it was Dr. Drysdale.
He had not long to wait. In 1861, Professor Lionel

Beale, of King's College, London, proclaimed as the
result of long-continued microscopical study that the
substance already recognised in plants and the lower
animals, and known as "sarcode" or "protoplasma,"
—that this structureless, transparent, colourless,
semi-fluid stuff, is universally diffused throughout the
human organism, and everywhere manifests distinctive vital characters, separating it from the "formed
material" around.   Into this it dies daily, just as the
whole organism dies into its *post mortem* condition,
when the mainspring of its common existence has
been broken.   But, till then, the "germinal matter,"
while ever perishing at its circumference, is ever re-
newing itself at its centre, converting the pabulum it
takes up into its own living substance, instinct and
quivering with vital motion.   All this Beale proved
as a fact which none could gainsay ; but the scientific
world at large was hardly ripe for the theory of life
which followed.   Drysdale, on the other hand, prepared by Fletcher, recognised it as that which he had
all along been holding, and saw in protoplasm the
very *materia vitæ* whose existence his master had
demonstrated, but which he had mistakenly identified.   At once, in the *British Journal of Homœopathy*,
he commenced a series of papers on the subject, cul-
minating in his separate publications entitled "Life
and the Equivalence of Force" and "The Protoplasmic Theory of Life." [1]   Here, while others were
hesitating, he boldly grasped and vindicated the doctrine ; and he had the gratification in 1879 of hearing

[1] Published by Bailliére.

it proclaimed as generally accepted from the Presidential chair of the British Association. Professor Allman there happily compared the place which protoplasm occupies in the mind of the biologist to that which the ether takes in the conceptions of the physicist. You know that, as sound consists of waves of air, transmitted from a vibrating body to our percipient organs, so light and heat — to go no farther — are best understood when regarded as undulations of some fluid medium. This cannot be the atmosphere, which extends only some forty miles above the earth's surface, while heat reaches us from the sun and light from the most distant of the fixed stars. We have therefore to assume the existence of an "ether," filling all interstellar space, as much finer than air as air is than water, and in like manner filling its "void and bare interstices." Protoplasm, as "co-extensive with the whole of organic nature — every vital act being referable to some mode or property of it — becomes to the biologist what the ether is to the physicist; only that instead of being a hypothetical conception, accepted as a reality only from its adequacy in the explanation of phenomena, it is a tangible and visible reality, which the chemist may analyse in his laboratory, the biologist scrutinise beneath his microscope and his dissecting-needle."

I am, I confess, desirous of pressing this conception upon you, on account of the part it must necessarily play in your thoughts about the system of Medicine you have adopted. Homœopathy is not a philosophy, but a method: yet it has a philosophy, and it is important that this should be clear and sound, and in harmony with the best thought of our time. Hitherto our notions of life have mostly been those enunciated by Hahnemann. In his student days it was always assumed to be a principle, distinct from the organism it animated, — the only question being whether Stahl should be followed in identifying it with, or Hoffmann in distinguishing it from, the immortal soul. When, in later life, he began to think upon the subject for himself, he espoused Hoffmann's doctrine, though with much more philosophic grasp. The vital principle with him was no entity, but a "force," and in conceiving it as such he anticipated (in 1829) the notion of force now everywhere current. "The organism" he wrote (in the edition of the *Organon* that year published) "is indeed the material machine to the life, but it is not conceivable without the animation imparted to it by the instinctively perceiving and regulating vital force, *and the vital force is not conceivable without the organism*, consequently the two together constitute a unity, although in thought our mind separates this unity into two distinct ideas, for the sake of facilitating the apprehension of it." The "vital force" is thus, with him, the mode of motion of the living organism.

Through Hahnemann such views of life have generally gained acceptance in the school which he has founded. They have, moreover, been reinforced among us from another and independent centre. In 1849 the distinguished French physician, Tessier, came over to homœopathy, and brought with him a brilliant

band of disciples. But he was more than a physician:
he too was a Master, and taught a complete system
of medical thought. His doctrine of life was that of
the Thomists, the followers of St. Thomas Aquinas.
It is fully set forth in the *Éléments de Pathologie* of
the present coryphæus of the school, Dr. Jousset —
one of the ornaments of our body. We are there
told that the cause of life, of that which makes living
beings distinct from others, is an "animating principle," which, existing already in the germ, preserves
it from corruption, fashions the organism into which
it grows and regulates all the functions of the same,
until its final retirement therefrom results in death
and dissolution. This animating principle also, like
Hahnemann's vital force, is said to be so intimately
united with the matter of which it is (in philosophic
language) the "form," that the two are inseparable,
— their union constituting the living being, one and
indivisible. But the Thomist doctrine has the dis-
advantage of lineal descent from that of Aristotle,
and adopts his terminology. The result is that it
speaks of the animating principle of a plant as the
vegetative *soul* (ψυχή, *anima*), and as it uses similar
language regarding the life of man, it seems to identify his soul with his vital force, and indeed to give
him one at all in no other sense but that in which a
cabbage has one. Moreover, if "anima forma cor-
poris" is to be the canon (and our *confrères* are so
fond of this phrase that for many years it has served
as a motto for their journal, *L'Art Médical*), — if
the soul is the form of the body, and we go to Aris-

totle to learn what is meant by "form," we find him
illustrating it by saying that vision is the form of the
eye.   The soul, then, so conceived, has not and cannot have any existence independent of the body of
which it is the function, — a conclusion tolerable
enough as regards animals and plants, but not to be
readily entertained by man, who "thinks he is not
made to die."   It is a conclusion, I should say, which
would be warmly repudiated by the school of which
I speak, which is distinctly Christian in its thought ;.
but it is one to which logic must remorselessly lead
them.[1]

Is such an "animating principle" required to account for the facts of the case ?   It is argued by
Dr. Frédault, another worthy inheritor of Tessier's
thoughts, that its existence alone will explain the
growth of the germ into the parental form.   But
crystallisation presents similar morphological phenomena, and no "âme minérale" is even supposed to
be working here.   It is perhaps easier to conceive of
death by calling this its "retirement:" but if it re-
tires, where does it go ?   Christian philosophy gives
an at least intelligible account of the future of the
disembodied spirit ; but what can be thought or said
of the separate existence of the bodily life — of the
bodily lives of the million creatures that perish every

---

[1] Mr. St. George Mivart, who represents in England the same Catholic
philosophy, would describe the ψυχή as a "principle of individuation," and
would grant its non-existence apart from the body.   But he does not grapple
with the difficulty in relation to man, save by suggesting that in his peculiar
case there may be some interference with the ordinary law (*Brit. Assoc. Report*, 1879.   Biological Section).

moment? Is it not much better to conceive of it as the "splendour and music" which "survive not the lamp and the lute"? Indeed, this last image, suggested by one of the interlocutors in the *Phædo* as representing the soul itself, and so casting doubt on its immortality, may well be applied to our present subject. Socrates (or Plato for him) has no difficulty in showing that the soul is more than the harmony of the body; but this is just what life is. It is a music, of which the organism is the instrument, and the natural stimuli the players, — secondary players indeed, for the great Musician who out of the diverse tones they elicit brings pure melody and perfect harmony is the Creator and Sustainer of all things, of whom Hahnemann was the constant worshipper, and whose reverent acknowledgment will I trust be always a feature of the school which calls him master.

You will observe that in all this I have been saying nothing of the distinctive faculties of man — of thought, and love, and conscience, and will. I have spoken only of the life which he has in common with the beasts that perish, aye, and with the plants of the field. His higher life may well have for its substratum some entity of another kind, some substance which we can only characterise by calling it spiritual. Hahnemann avoided the dilemma of the Thomists by distinguishing the vital force from the rational and immortal soul. But his conception of life as a "force," though philosophically tenable, lacks scientific warrant. The forces are affections of matter, modes of its motion, correlative and interchangeable

among themselves, and amenable as matter itself to
the law, "omnia mutantur, nil interit." But matter has
other qualities besides these — often casual — affections : it has inherent *properties*, which make it and
its various forms what each essentially is. According to its properties it is influenced by the natural
forces : its chemical affinities determine how it shall
act under chemical attraction, its responsivity to
magnetism makes it advance towards a magnet when
presented to it. Is life, then, a force or a property?
Clearly the latter. Vitality is, as I have said, the
living matter's capability of response to the natural
stimuli ; and the life of each individual is that response on its part, conditioned by its own special
constitution. Some of these stimuli — as light and
heat — are themselves forces, and the work done
under their influence is of course an expenditure of
energy equivalent to that which they convey. But
to dead organic matter they appeal in vain, — not
that they are less stimulating, but that it has no
longer the capacity of answering to the prick of the
spur.

The importance of this conception in our philosophy of homœopathy is that it keeps us clear of meta-
physics, that it enables us to be dynamists without
being spiritists. Hahnemann did not always maintain the lucid idea of the vital force whose expression
I have read you from the *Organon*. He frequently
speaks of it as something distinct from the organism
it inhabits — an *archæus* regulating its development
and motions. Revolted by the gross humoralism

of his day, he retreated beyond solidism into pneu-
matism : he taught that all disease originated in a
derangement of the vital force, physical changes
being only secondary.   The exciting causes of disease, especially the infectious principles, must therefore be non-material ; and in like manner the drugs
which modify it for good must be so attenuated as
to part with their physical envelope, and become
something new and spirit-like.   I am not prejudicing
the question of dose : I am rather protesting against
its being prejudged by the assumption that the seat
of medicinal action is an invisible, intangible "force."
Recognise that it is living matter — protoplasm ; and
then you have all that dynamism can desire, without
the need of any such untenable theory as "dynami-
sation."   You have matter in its most sensitive,
responsive state, capable of modification by the finest
germs or molecules ; but it is matter still, requiring
its morbific *contagia*, its sanative medicines, to be —
however minute — material also.   This Science also
requires, for (in the natural world) she knows of no
force apart from matter, none that is not resolvable
into the motion originally impressed upon the atoms
of the Universe by their Creator.   The amount of
energy contained in and put forth by matter varies
indefinitely, and we avail ourselves of the variations ;
but to think that we can escape from the matter and
retain the energy is a delusive dream.   I hope that
no student of Boston University, of the Medical
School which has Conrad Wesselhoeft among its
Professors, will ever dream it.

II.

THE KNOWLEDGE OF HEALTH.

THE subject of medical art being the human organism, there are two things which the physician has to do for it : he has to preserve it in health, and — when necessary — to free it from disease. In the execution of the first of these tasks he is a hygienist, in the second a therapeutist. It is of the health-preserving duties of the physician that I am to speak to-day. In allotting no small space to this subject I am but treading in the footsteps of the master of our school. It is sometimes ignorantly represented that Hahnemann was a mere drug-giver, that — as Dr. Bristowe has said — "for him, preventive medicine, which deals specially with the causes of disease, and has been successful only in proportion to its knowledge of them, would have been a mockery and a snare." So far is this from the truth, that in his first homœopathic essay of 1796 Hahnemann speaks of the removal of causes as the most elevated way which practical medicine can follow, as a "royal road;" and in his *Organon*, from its first edition in 1810 to its last in 1833, he begins by assuming that both to pre-

22

vent disease, and to make his curative treatment
unobstructed and permanent, the physician will also
be a hygienist. These precepts are well illustrated
by his own example, as seen in his letters of advice
to patients extant here and there in our literature.
He, indeed, who anticipated Pinel in a rational and
humane treatment of lunatics, was not likely to have
neglected the general management of his sane patients ; and his writings prior to the time of his special
occupation with his therapeutic method frequently
touch on the subject. In so acting he has been followed by his most genuine expositor on American
soil — the late Carroll Dunham ; who, in his "Homœopathy the Science of Therapeutics," begins by defin-
ing the sphere and emphasising the importance of
hygiene. In the same spirit have many of the homœopathic physicians of the United States taken active
part in the Public Health Associations of their country ; and one of the foremost European hygienists —
Dr. Mathias Roth — belongs to the homœopathic
body in England.

It could hardly, indeed, be otherwise ; for the
method of Hahnemann in its very nature links itself
with healthy living. Its single, clean, mostly taste-
less and inodorous remedies suggest a regimen equally
simple and pure : the gentleness of its medication
consists only with a like non-perturbative dietary.
He that dispenses with narcotics and blisters in illness will hardly abuse stimulants and condiments in
health. The importance we attach to drug-giving
makes us the more anxious to banish all that is medi-

cinal — that is, poisonous — from our patient's surroundings, and so to secure for him the prime necessities of health — fresh air, pure water, and wholesome food. We are thus hygienists in virtue of our method ; and still more because we are not homœopathists only, but physicians, and — *nihil de humano corpore a nobis alienum putamus.*

Milton has said that a poet's life should itself be a poem ; and I would apply this saying to the physician. He who preaches healthy living should himself practise and illustrate it. With what force can he enjoin moderation at table if he is a *bon vivant?* how can he warn his patients against nicotinic intoxication if he smokes all day ? And more, — he must see that his very virtues do not lead him to excess : he must restrain himself from the noble folly of overwork, that he may with greater effect reprove it in others. He ought to be able to say, — by the observance of the principles I commend to you I am myself a healthy man, able to do a full life's work with strength and joy. Of course there must be exceptions to such a statement — cases where an unhealthy tendency has been inherited, or accident has invaded and broken the constitution. But even here much may be done : if it could not be, how could we expect patients similarly circumstanced to place themselves in our hands ? When all allowances have been made, it remains true that the physician should exemplify the ideal he holds up, and lead as well as point the way to health.

This is the first duty I would lay upon you in connexion with hygiene ; and the next is that you should have clear conceptions as to its constituent elements. Hygiene may be defined in brief as man's due care for himself and proper selection of his environment. His body needs exercise, rest, warmth, air and food. The first two it can give itself, and the third it can to some extent supply ; but air to breathe and food to eat come to it from without, and health largely depends on the quantity and quality in which they are furnished to it. Let me say a few words on each of these heads.

1. The recommendation of *exercise* is a time-honored function of the physician, and probably in Hippocratic times it was prescribed with as much minuteness as medicines are at present. It forms now as ever a prime necessity of health and adjuvant to recovery ; and to its hitherto-known varieties of walking, riding, driving, swimming, and gymnastics, we can add the several kinds of "cycling" and the "massage." Let us think, however, why it is that exercise is desirable. It is so because the more rapidly the effete portions of our organism can be carried off, and replaced with new matter, the higher our health and efficiency. Exercise promotes this moulting, by quickening the blood-current and increasing the secretions ; while the waste thus induced causes appetite — a demand for food to supply it, and thus deposition of fresh tissue. The enormous quantity of nutriment which can be taken and digested by the subjects of Dr. Weir Mitchell's treatment, while undergoing vigorous massage, well illustrates this point. One sees, therefore, on what class

of persons exercise should be enjoined.   The young
hardly need the admonition, — the problem of per-
petual motion going far towards solution with them ;
and the old should not receive it, for, their power of
repair being feeble, it is unadvisable to precipitate
waste.   It is the middle-aged — embracing in this
category all lives from twenty to sixty ; and among
them those of luxurious or studious habits, and those
whose occupations are sedentary.   Such persons
may with little exercise strike a sort of secondary
balance — may compensate insufficient outflow with
moderate in-put, and so establish a working equilibrium ; but they know little of the sensation of puis-
sant life and the buoyant energy for work which
comes of active exertion and hearty feeding.

2. In the present day, perhaps, *rest* has to be enjoined more frequently than exercise.   It is an age
of overwork ; and nothing is more common than to
hear, both in public and in private life, that such and
such an one has been ordered by his physicians to abstain entirely from work for a longer or shorter time.
You must of course not shrink from giving such advice when it is necessary.   But on the one hand I
would urge upon you not lightly to lay such an obliga-
tion on your patients ; to put yourself in their place
before you do so, and realise the serious consequences
it involves.   I do not know how it is on this side of
the Atlantic ; but on ours it is becoming quite a fash-
ion to "order away" every troublesome case on which
a man is consulted, without considering what this
means for the sufferer, and instead of taking pains to

see whether by medication and wise regulation of life
so revolutionary a change might not be averted.   This
I say on the one hand ; and on the other I would commend to you the cultivation of preventive medicine
here.   The physician should preach — in season and
out of season, he should also practise and exemplify,
the place of rest in life.   I have called overwork a
noble folly,. and sometimes indeed the adjective has
more truth in it than the substantive.   Of one of our
ancient kings — and here the " our " belongs alike to
you and to me, to the Anglo-Saxon race throughout
the world ; of Alfred the Great it is related that he
divided his day into three parts, giving eight hours
to the cares of state, eight to study and writing, and
allowing only the remaining eight for repose, recrea-
tion, exercise and meals.   It was perhaps needful for
him so to expend himself, and the nobleness of his
devotion has enshrined him in the reverence of all
succeeding ages ; but it is little wonder that for most
of his life he was racked with painful disease, and
that he died worn out at fifty-two.   I advise no one
for lesser cause, and *à fortiori* not for ambition or gain,
to allow so little breathing-space in life for rest and
leisure.   To a man in active work eight hours are
none too much for sleep alone ; and if another eight
are allowed for the care of the body and the play of
the lighter faculties I am sure that the work done in
the time which remains will be all the better in quality.   It will be this, and it will go on longer.   There
will not be the tragedy so often enacted in these days,
of a man in the prime of his powers and usefulness

more or less suddenly "breaking down," all his accumulated force and experience lost to the world
because he had not husbanded his strength. This
wisdom must be preached; and so must that be which
lifts a warning against the feverish haste, the over-
pressure, of our present mode of existence. Men
try to make up for it by taking long holidays; but
these — even if themselves not spent in hurrying
through foreign sights and sensations — do not make
up for the lack of daily repose. The heart, that tire-
less worker, sets us an example here, for it rests
during a fourth part of every pulsation, and beats at
an almost uniform rate all the day long. Why should
not we work thus evenly, rest at least thus proportionately?

3. About *warmth*, I apprehend the most important truth to be that the body makes its own heat,
has its hearth and furnace within itself. How few
people realise that the main use of clothing is to keep
in the caloric which the frame has generated — any
end it may subserve in protecting from draughts of
air being quite subsidiary! In choosing warm clothing, therefore, it is not so much quantity or thickness
we have to study, as that the material shall be non-
conductive of heat, and its fit such as to allow space
for a stratum of warm air near the body. And, even
here, it ought to be borne in mind that it is the *sensation* of warmth we are conserving rather than its
actual existence. Nature has so bountifully provided
for us, that no change in environment makes much
difference in our internal temperature, — cold to the

surface only stimulating heat-production, as a fervid atmosphere is compensated by increased heat-radiation. Thus the normal warmth of the body — that which is necessary for the due performance of the vital processes — rarely suffers loss, so long as fuel can be supplied in food, and combustion kept up by exercise. That which fluctuates is the sense of warmth, which is due to the state of the tissue in which the sensory nerves of temperature terminate — the skin. If the cutaneous arterioles are sufficiently open, and the heart pumps the blood freely through them, we feel warm, and so are comfortable. Now life without comfort is hardly worth living, and so the sense of warmth is to be cultivated. Sometimes its absence is a real morbid condition, and we have medicines to help us here, — drugs like Asarum and Ledum, which make healthy persons chilly. But more commonly the deficiency is a matter of hygiene, and can be supplied by improving the state of the skin. Foremost among the measures available here is that which we English affectionately call our "morning tub" — the cold bath taken immediately on rising. I speak not of it now as it conduces to cleanliness, — for this may be otherwise secured. I am regarding it as a means of making the skin less sensitive to external cold, — of approximating the whole surface to the callousness of the face, which we expose to the coldest blasts without distress or peril. I am thinking also of the glow which follows, and which is so readily renewed by food, and exercise, and warmth from without. Every man or woman who can take a cold bath in the morning should have it, and will be the happier and the healthier for it.

But there are cases in which heat-production is absolutely below the normal standard, and vitality is lowered in proportion. Here again we may have a true — that is, primary — morbid condition, and specific medication may come to our aid. One of the most interesting things in Dr. Burnett's essay, " Natrum muriaticum " (which I trust you have all read, or will read), is the effect of that remedy in promoting the calorifacient function. Calcarea carbonica probably has a similar action, though its coldness is rather felt locally — as in the head or legs — than generally. But here also the defect is more commonly a physiological one, and amenable to hygienic measures. We have only to think over the sources of animal heat to see how much may be done by judicious regimen to increase its production, as by urging exercise — active or passive, and supplying readily-oxidised food. At the two extremes of life, indeed, our aim is somewhat different. In the very young heat is formed freely enough, but too readily lost ; and you can hardly be too earnest with mothers that they should clothe their children warmly, and not sacrifice health and comfort to appearances. In the aged, little can be done for the making of heat : it is its supply from without to which we must attend. Warmth to the surface, warm air to breathe, are our adjuvants here ; and especially helpful are hot drinks, which add positive caloric to the system, and — for the time at least — inspirit all its functions.

4. That the *air* we breathe should be pure is perhaps the best-learned lesson of hygienic science. All educated persons recognise the truth, and nearly all act up to it so far as supply of oxygen and elimination of carbonic acid are concerned. But our modern sanitary arrangements, with all their advantages, have brought into being a new agent wherewith to contaminate the atmosphere : I refer to sewer-gas. I am sure that we cannot be too much on the alert to detect the possible existence of this *causa mali ;* and the disconnection of drain-pipes will often do far more for our patients' health — to say nothing of our own reputation — than the most carefully selected medicines. A good instance of the subtle way in which sewer-gas will act is supplied by our recent literature. In the *Monthly Homœopathic Review* for January, 1879, Dr. Edward Madden related a case of erysipelas occurring in a child of his after vaccination with calf-lymph. His inference was that the use of this matter does not secure immunity from such accidents. In the April number, however, appeared a letter from Dr. Edward Blake asking Dr. Madden if he was sure about the state of his drainage, as he had lately traced a post-vaccination erysipelas to a fault in this direction. In August Dr. Madden replied, saying that investigation had shown his house to be "exposed to nearly all the dangers of admitting sewer-gas to which defective arrangements could conduce." Dr. Blake has indeed made the subject his own, and supplies another instance of a member of our body being an earnest hygienist. In

his brochure entitled "Sewage Poisoning : its causes and cure,"[1] he shows how perilous is this taint in the air to the subject of an operation or the parturient woman, and traces to its influence many forms of ill-health, as diphtheritic throats, morning headache and diarrhœa, unaccountable languor, anorexia, feverishness, and sleeplessness, passing on — especially in children — to anæmia and glandular suppurations. To these probably every man's experience would supply additions, — I, for example, having seen ascarides in an adult obstinately persist while it was present, and immediately clear away on its removal. Sewergas cannot of itself initiate a specific disease, but it both spreads its germs and predisposes to their reception, and then — when the malady has been set up — tends to make it malignant, and — if the patient recovers —to retard his convalescence. Whenever there is an x in the morbid conditions with which you have to deal — some unknown factor making everything go badly and defeating your most just anticipations, there suspect sewer-gas, and look to the drainage.

The air we breathe must be pure, and it must be suited to any weakness — any vulnerability which may exist in our constitutions. In health "man is man, and master of his fate :" he can subdue to his purposes the greatest extremes of climate as he can the beast of the field, the fowl of the air, and whatsoever passeth through the paths of the seas. But let his system once give way at any point, and his

[1] Spon ; 16 Charing Cross, London, and 446 Broome Street, New York.

mastery is over : *non vincit*, now, *sed parcundo*. He must find the spot of earth best suited to his *pars minoris resistentiæ*, if the rest of his frame is to flourish unimpaired. This choice of climate fre-quently falls to the physician to determine, and we must be prepared to advise alike with sound princi-ples and with knowledge of their practical applica-tion. There is nothing in which we need so specially to individualise as this. The total which we call climate is the product of several factors, — latitude, exposure, elevation, soil, water, all go to make it up ; and it is the whole which we must estimate and fit. Nothing is more instructive than the maps of the British Islands which Dr. Haviland has framed, showing by diverse colouring the tracts more or less favourable to the development of consumption and cancer. Inductive generalisation can perhaps estab-lish some common features here — defective flow of air in the one case, saturation of soil in the other ; but for the present the individual facts are our most im-portant knowledge. Again, what is more significant than the fact elicited by Dr. Chambers, that in Italy chronic degenerative maladies like Bright's disease play a very small part among the causes of mortality compared with Great Britain? Residence there, when practicable, is obviously the best resource for those so afflicted. The value of mountain-air, though cold, for phthisical subjects is another fact of the same order. You doubtless have, in the vast extent of your States, an England and an Italy ; a Riviera or Madeira for your delicate-chested, a Davos for your

true consumptives. Study them individually, as we
have had to study the health-resorts of Europe ; and
you will have in climate a most potent agent at your
command.

5. We come now to the important subject of *food*,
and with this must occupy ourselves for the remainder of the present lecture.

Why must we eat ? That nature prompts us so
to do by the calls of appetite, and sternly punishes
any failure to attend to her admonitions, is obvious
enough : but what does science say on the subject ?
for what end or ends do we take food ? This question is (I think) best answered by the old comparison
of the body to a steam-engine, whose fire is vitality,
and which needs fuel that it may perform work. The
human engine differs indeed from its mechanical
type that it has, at one stage of its existence, to
grow, and throughout its continuance to repair its
own waste. For these purposes food is required as
substance, to make the tissues which need augment-
ing or replenishing ; and until lately this (with the
supply of heat) was looked upon as its sole office.
Now, however, we have come to see that in the adult
body there is not much waste ; that its floating cap-
ital is small in proportion to its actual assets. If,
then, repair were the sole use of food when supply
for growth was no longer needed, the demand for it
would be very small. But the doctrine of force in
modern science has taught us that for all work done
energy must be expended, and so — if the work is to
continue in doing — must be renewed. For its sup-

ply the main source is the aliment. I doubt not that this is (mostly, for there are exceptions) assimilated before it is oxidised — that it is its own substance which the animal steam-engine burns, and not directly the coal. But so rapid is the conversion, so unstable the product, that the pabulum may fairly be regarded as the fuel for the flame, and the immediate source of the motive-power of the body.

" The human body" says Hermann, "like that of every other animal, is an organism in which, by the chemical changes of its constituent parts, *potential* is converted into *kinetic* energy. . . . These chemical changes depend on the presence, within the organism, of energy-yielding substances. . . . Every act of the organism must diminish, to a corresponding extent, the energy-yielding store which it contains. . . . It is, therefore, essential, in order that the organism should continue to exist, that it be continuously supplied with oxygen and oxidisable substances. The latter are called the organic constituents of food."

He goes on to show that the inorganic elements of our diet, besides any mechanical (*e.g.* solvent) uses they subserve, supply the waste of the same substances which form part of the body; but do not furnish it with energy.

Even stated thus, it appears that food is the fuel for the flame of life. But we may go farther when we regard those dialytic processes which are not oxidations, and in which aliments are decomposed immediately without previous assimilation. What occurs when sugar — or, what ultimately comes to the same thing, starch — is ingested? It is converted.

into lactic acid, a substance of much lower equivalent value; and thus energy is directly evolved   A similar thing happens with fats, where — mainly under the influence of the pancreatic juice — the stearine (or what corresponds to it) is broken up into stearic acid and glycerine; and this prior to assimilation.   Here are liberations of force occurring in the food itself; and when we consider how large a part is played in our ordinary diet by sugar, starch and fat, the conception of it as fuel becomes yet more strictly warranted.

An important practical conclusion, moreover, follows in respect to the quality of our food.   Of old, when our whole body was supposed to be in a state of continual flux, and the main use of food was conceived to be the laying down of tissue, it was natural to look upon flesh as flesh's best repairer.   The earlier achievements of organic chemistry favored the same result, for in all physical exertion great waste of muscular tissue was supposed to occur, which meat diet alone could replace.   But, as Dr. Pavy justly says,[1] "the information which has been obtained during the last few years has completely revolutionised some of the cardinal scientific notions formerly entertained" on this subject.   It is now ascertained that the most active and prolonged muscular exercise causes little increase in the excretion of urea, and therefore as little loss of nitrogen; while the capacities of herbivorous animals and insects find ready parallels in the human race.   "The inhabitants of mountainous dis-

[1] Treatise on Food and Dietetics.   1874.

tricts prefer to take fat and sugar as provisions when they have arduous journeys to perform (Hermann); and Dr. Anna Kingsford [1] has well shown that a great deal of the most vigorous work of the world is done by men in whose diet flesh plays little or no part.

Shall we, then, follow her to her conclusion that we should all be vegetarians? I think not. I quite acknowledge that man is essentially a frugivorous and graminivorous animal; and that on fruit, pulse, and cereals, with milk and water to drink, he may keep himself in health and vigour. But man is something more than an animal; and as he has gone beyond his simian fellows (I will not say ancestors) in learning to cook his food, so has he in widening the range of his diet. He has done for himself what he does for his domestic animals, teaching dogs and cats to live on bread and milk, which their wild brothers — wolves and tigers —would starve rather than touch. The Biblical tradition of permission to eat flesh having first been given to Noah after the Deluge, — primitive man having been a vegetarian,[2] suggests such aliment as an advance on his part, suited to change of environment; and such it assuredly is, for in cold and damp climates animal food can rarely be dispensed with. Chemistry witnesses to its usefulness, for it shows a certain amount of nitrogen as necessary to be supplied in the diet; and meat gives us this in smaller compass than any other food, and with greater relish. There is, moreover, a stimulant

[1] The Perfect Way in Diet. Lond., Kegan Paul. 1881.
[2] Genesis ix. 3; I. 29.

quality about meat — especially roast meat — which no other food can supply. The vegetarians object to this, saying that it gives to man some of the ferocity of the carnivora. But a slight infusion of their characteristic nature is possibly not without benefit to the human stock, doing for it what the blending of Norman with Saxon did for the Englishman.

But while I cannot endorse or practise, in its entirety, the vegetarian regimen, I do so far agree with its advocates that flesh should be the exception, instead of the rule, in human dietary. The three meat meals a day, so favoured in my country (I do not know how it is with you), are alike wasteful and injurious. Wasteful, because animal food — fish, flesh, or fowl — is the most costly of our aliments; and injurious, because less force can be made out of it in proportion to its bulk than out of any other article of diet, and the residual substance goes to burden the liver with the formation and the kidneys with the elimination of urea to an extent quite beyond their powers. Hence — unless the most vigorous out-door exercise counteract the evil — there comes lithæmia, azoturia, and gout, if not organic disease itself. I am of course speaking of the lean of meat: the fat occupies a different place, and its ingestion need be limited only by the powers of digestion. Bacon or eggs thus suggest themselves as the most suitable adjunct to breakfast, and cheese to lunch or supper, where we do not care to make these purely vegetable meals, confining the consumption of lean meat to dinner only.

I would sum up the leading principles of diet thus : —

a. The growing child requires food for the supply of substance, and cannot have too much, if he can digest it. Milk, oat- and wheat- meal, sugar and fruits are his main subsistence until he begins to learn, and then meat should be added.

b. The adult requires a certain small amount of nitrogenous food to repair waste, and this he can best get in meat. But his main requirement is force for work, which can be derived from carbo-hydrates and hydro-carbons — starches, sugars and fats — with less expenditure than from any other source. The Irishman gets these in his potatoes and buttermilk, the Hindoo in his rice and ghee ; and either needs only a small proportion of meat or pulse to raise him to full vigour.

So far as to what we should eat ; and now as to what our drink should be. First of all I would say that few people drink enough water. They forget that this element constitutes at least two-thirds — some say four-fifths — of our frame, so that it is a veritable food, especially as no part of us is being more constantly given off. Your plucky compatriot, Dr. Tanner, lately interested the whole civilised world by showing how a man could live for forty days on water alone ; and he was only illustrating thereby to an unusual degree a recognised scientific truth. Water is further a diluent and solvent, and as such plays a most important part in the system, — all whose processes languish without it. A pure water-supply is the prime necessity of every habitation, — not only as excluding germs of disease, but as encouraging the free imbibition of the fluid. The cautions sometimes given against it are mere fancies. You are told that you must not drink of the mountain streams while pedestrianising in Switzerland : I have more than once done so copiously with perfect impunity. I was warned when I visited your Philadelphia in the "heated term" of 1876 that I had better not make too much use of ice-water; but I should be sorry to say how many gallons my thirsty lips disposed of with nothing but refreshment.

And now what shall we say to the tea, coffee and cocoa we drink warm at some of our meals? The cocoa we may dismiss as little more than a liquid food — one, indeed, of the utmost value, but not concerning us here. Tea and coffee occupy an identical place, save that the latter is a little more heating. Both diminish waste of tissue, both stimulate the nervous centres. They are therefore admirably adapted to circumstances involving much exposure or fatigue ; and we can well understand how the Arctic voyager prizes them, and what store the Peruvian mountaineer sets on his analogous coca. But, *per contra*, they seem less suitable to normal conditions and healthy bodies ; and, though we may not condemn coffee so strongly as Hahnemann did, or anathematise tea with the Dean of Bangor, I think there can be no doubt of their frequent abuse. I would keep them entirely away from the young ; and if adults must use them, would have them ordinarily made too weak to produce any appreciable effect, increasing strength only according to need. Even then we must beware of being lulled to false security by their potent action. They check waste of tissue, and so far are useful as temporary conservators; but we must remember that the material thus retained is partially effete, and that true health lies in its removal and replacement. By their gentle stimulation they take off the sense of fatigue; but this sense is nature's cry for repose, and if it is hushed up instead of its appeal being granted, a nemesis will sooner or later follow.

I may here speak of tobacco, which, though not taken through the digestive organs, yet behaves in the system like tea and coffee, so far as the tissues are concerned. It thus partakes of their advantages and of the objections to them; and, like them, should be used — if at all — in the strictest moderation. We do not drink tea and coffee all day, but so — I fear — do many smoke tobacco. If you ask them why they thus act, they will tell you that they enjoy the soothing effect on the nervous system. Tobacco differs indeed from the substances I have just mentioned in being a sedative here, not a stimulant. It depresses from the first, and its ultimate action is prostration and paralysis; while tea and coffee, carried to excess, over-excite. Well: if the brain and nerves are temporarily disturbed, and want quieting, I would not grudge the solace which a pipe can bestow. But surely this does not occur to all men every day, and several times in a day, as would be inferred from the

practice of most smokers. To induce sedation without cause is — to speak plainly — a slight self-poison-
ing ; and those who incur it should realise what they
are doing.

Last, of alcohol. I have nothing to say here on
the moral aspect of "drinking" — of its peril of inducing intoxication, which we all agree to condemn.
Nor would I speak of the usefulness of alcohol as a
medicine, which we all — or nearly all — agree in ac-
knowledging. The question is as to its place in
health. Is it a food? The advocates of total absti-
nence maintain the contrary, and base their denial
on the experiments of Lallemand, Duroy and Perrin,
which seemed to prove that it is entirely eliminated
by the emunctories. But they ignore the counter-
experiments of Anstie and Dupré, which — after
"some fourteen years of almost unintermittent work"
— have established the fact that a non-intoxicating
dose of alcohol is almost entirely consumed within
the body,[1] and that within a very short time. Oxidation is the only conceivable mode of its destruction, and this implies the liberation of force, which,
as it does not appear as heat — alcohol actually lower-
ing the temperature, must become working energy.
Alcohol therefore is certainly a force-producing food ;
and many instances are on record of life being sustained almost wholly upon it for months and even
years. It is thus, probably, that it acts as a "stimulant," for its medicinal influence is quite in the opposite direction. The earliest manifestation of its in-

[1] See *Practitioner*, xiii. 15.

fluence is flushing of the surface and quickening of
the heart's action ; precisely parallel to that produced
by the inhalation of nitrite of amyl, and like that
associated with diminished arterial tension and fall
of internal temperature (despite the illusive sense of
warmth caused by the glow on the skin), and so
pretty certainly owning a similar causation, viz. : vaso-
motor paralysis. The further progress of its action
is to depress the other parts of the nervous system,
beginning with the centres of speech and motion,
and ending with those of consciousness. It is also
an irritant to all tissues with which it comes in contact, either locally or by absorption ; and shares
with tea, coffee, and tobacco the power of checking waste of substance. These undesirable collateral effects spoil it for a food under ordinary
circumstances, though it may be made such for
special times — as during exhausting diseases. A
healthy man may drink a tumbler of light beer or
a glass of pure wine at his dinner with impunity.
But if he take the stronger, fortified wines and malt
liquors, and the spirits, and in such quantity as to
feel appreciable effects from them, he is inflicting on
himself the worst of slow poisoning. He is diminishing his resistance to cold and his production of
nervous energy : he is fretting· stomach, liver, kidneys, brain, into smouldering irritation ; and he is
storing effete tissue which will break down under the
least provocation, as seen in brewers' draymen when
they come into hospital from illness or accident.

The conclusion must be that the use of alcoholic

drinks in health is a custom more honoured in the breach than in the observance. It is, for most people, entirely unnecessary; and, when practised, should be so on the most limited scale. A man should be independent of it; and, that he may be so, should be brought up without it. Let the dietary of the young be non-alcoholic, if you love them; and then, as they grow up, you may pretty safely leave them to do as they like.

I fear I may have wearied you, ladies and gentlemen, with these details, but their introduction was my only mode of doing justice to the subject. They are not taken merely from books; but represent the matured results of nearly thirty years' study and observation of their phenomena. I cannot but indulge the hope that they may be helpful to my younger hearers, in enhancing — for their own benefit and their patients' — their knowledge of health.

III.

THE KNOWLEDGE OF DISEASE.

THE physician is a hygienist, but he is above all things a healer. His "high and sole mission" as Hahnemann states at the outset of the *Organon* "is to restore the sick to health — to cure, as we term it." In this pursuit he has to deal with *disease*, as the subject of his operations, and with *drugs*, as the chief instruments with which he works. The greater part of your medical life, whether as students or as practitioners, will have to be spent in acquiring and applying the knowledge of these two. It cannot, therefore, be without profit that we should spend some time in considering what such knowledge should be — in what proportion it should deal with the phenomena, the laws, and the causes, which we have seen to be the elements of all knowing.

I suppose that all lecturers on the Practice of Physic begin the account of particular diseases by describing their clinical features. "Every now and then," as my former teacher at King's College — Dr. George Budd — used to say, "we meet with" cases

45

presenting such and such groups of phenomena and
sensations. He would then give the name by which
the malady thus constituted is styled, and would proceed to relate how it came about, and wherein it
essentially consisted, so far as these points were
known. But observe the difference involved in this
" so far." The ætiology and pathology of the disease
were more or less uncertain, and our conception of
them was liable to vary as new facts came into our
view. But its clinical features remained. They
were those which perchance Sydenham, or even Hippocrates, had described as graphically as any modern
physician : they, amid all shiftings of conception
about them, were permanent and sure.

Now when Hahnemann proposed as the law of
cure *similia similibus curentur,* " let likes be treated
by likes," and had to define the elements of the comparison thus implied, he took these features as the
disease-basis of his method. Simplicity and certainty
were his aims in practical medicine. He could not
conceive that the obstacles to them were insurmount-
able ;[1] and he felt sure — in his profound Theism —
that " as the wise and beneficent Creator has per-
mitted those innumerable states of the human body
differing from health, which we term disease, He
must at the same time have revealed to us a distinct
mode whereby we may obtain a knowledge of diseases,
that shall suffice to enable us to employ the remedies

---

[1] I refer to his essay published in Hufeland's Journal in 1797, entitled " Are
the Obstacles to Simplicity and Certainty in Practical Medicine insurmount-
able ? " (Lesser Writings, p. 358.)

capable of subduing them, . . . if He did not mean
to leave His children helpless, or to require of them
what was beyond their power." This "distinct
mode" was, he considered, the clinical. He was indeed far from refusing the aid of ætiology, to such
extent as it was available. In the *Organon* he points
out that it is obviously part of the physician's duty
to ascertain the presence or incidence of any exciting
causes of disease, that he may remove them now and
ensure their avoidance in future. It is also desirable,
according to his teaching, to discover the past causes
— both predisposing and exciting — of the patient's
morbid condition, as certain medicines are found specially suitable when disease has originated in certain
ways — Arnica when from injuries, Rhus and Dulcamara when from damp, and so forth. Pathology,
however, Hahnemann absolutely rejected for therapeutic purposes. It was in his day far more a matter
of guess-work than it is now ; and was too much of a
quicksand for a sure foundation to be laid in it. But
he went farther, and maintained that a knowledge of
the essential nature of disease was both unattainable
and useless. His views on this subject are best expressed in § 5 and 6 of the fourth edition of the *Organon*. "It may be conceived" he there writes "that
every disease implies a change in the interior of the
human organism. But this change can only be *inferred* by the understanding, and that in a dim, misleading manner, from the morbid signs present. It
is in itself unknowable, and perhaps in no way without illusion can the nature of the internal invisible

change be apprehended.   The invisible morbid alteration in the interior, and the outward change of condition noticeable by our senses (the totality of the
symptoms), together constitute, to the gaze of the
Almighty Creator, that which we call disease.   But
the sum of the symptoms is the only side of the malady accessible to the physician, the only one observ-
able by him, and the principal thing which he can
learn from the patient and employ for his benefit."
The side of disease which pathology explores was
thus to Hahnemann its *noumenon* in the strict sense
of the word, — recognised metaphysically as existent,
but taken no practical account of; for all purposes
but those of thought represented by the phenomena.
The "totality of the symptoms" is, to the therapeutist, the disease.

Is this position tenable?   Most persons would at
once answer in the negative; but they would do so,
I think, without regard to the end set before us in
thus limiting our apprehension of disease.   If we
were dealing with it as an object of science, a branch
of natural history, it is certain that symptomatology
would be an insufficient basis for our knowledge.
No one has better shewn this than Liebermeister,
in his Introduction to the section on Infectious Diseases in Ziemssen's *Cyclopædia*.   The basing unities
of disease on symptoms gave us such pathological
entities as hydrops, icterus, apoplexy, and the like;
and "from this symptomatic stand-point quotidian
fever was a different malady from the tertian or the
quartan form, while on the other hand ascites and

tympanites were only different forms of the same disease." He goes on to argue that the most scientific — because most real — rule of classification must refer to *causes*, must be ætiological. The quotidian and quartan types of fever are one, because they both originate from malaria : they are to be differentiated from pyæmic febrile attacks, though these may have a similar rhythm and similar symptoms, but are to have grouped with them other malarial affections which differ greatly in symptoms, such as malarial neuralgia, malarial diarrhœa, malarial cachexia. "The lightest form of varioloid is regarded as essentially identical with the most severe form of variola : on the other hand, vaccinia and varicella are separated from it. . . . The simplest diarrhœa, arising from the poison of Asiatic cholera, is to be ascribed to this disease : on the other hand, a very severe and deadly cholera morbus is to be marked as another malady."

Nor is it for classification only that such scientific knowledge of morbid states can be turned to account. It avails for prognosis. That we are able to distinguish true typhus from other forms of continued fever, and that we know its natural history, enables us to affirm that if the patient passes the nadir of his prostration between the fourteenth and seventeenth day, and then displays an upward tendency, he will pretty certainly recover. It avails for the general management of the patient. To recognise relapsing fever as present leads to a care being taken after the first apparent recovery which would

otherwise be needless, but which here materially
influences the course of the second paroxysm : it
also suggests the use of antiseptics during the inter-
val for the possible prevention of the recurrence, as
was carried out so successfully by Dr. Dyce Brown
in Aberdeen.[1] It avails, again, for estimating the
influence of treatment.   Of old, every chancre which
disappeared without secondaries supervening went
to the credit of the mercury that was given, or to
the demonstration of its needlessness if it had been
omitted.   We now know that the soft chancre —
which occurs by far the more frequently of the two
— has no such significance, and is naturally without
sequelæ.

Now if medicine were an applied science only, it
would be with such knowledge and its utilisation
entirely that we should be concerned.   But it is the
merit of homœopathy that in it medicine assumes its
true place in being an art — the art of healing.   The
physician is not primarily a cultivator of science : he
is a craftsman, the practiser of an art, and skill
rather than knowledge is his qualification.   His art,
indeed, like all others, has its associated sciences.
Physiology and pathology are to it what chemistry
is to agriculture, and astronomy to navigation.   So
far as they bring real knowledge, the more versed
the physician is in them the better for himself and
for those in whose aid he works.   But he was before
they had their being, and his art should have a life
of its own independent of the nourishment they

[1] See *Brit. Journ. of Hom.,* xxxi. 355.

bring. They must, being progressive, consist largely
of uncertainties — working hypotheses and imperfect generalisations, destined ere long to be superseded by more authentic conceptions. Medicine
should not vary with their fluctuations, or hold its
maxims at the mercy of their support. While grateful for the aid they bring, it should go on its own
separate way and fulfil its distinctive mission. The
method of Hahnemann enables it to do this, by taking the clinical aspect of disease as its working
basis. Pathological knowledge has little to do with
drug-selection so determined. It has taught us —
for instance — to recognise enteric fever as specifically distinct from typhus, and for many purposes
this differentiation is highly important. But the
indications for its homœopathic remedies were just
as plain when it was classed merely as "typhus abdominalis," and were as well given of old by Wolf
and Trinks as they now are by Jousset and Panelli.

Again, if our aim be the ascertainment of the
particular organ affected in a given case, symptomatology is certainly insufficient. Not, indeed, because
it is to be distinguished from physical diagnosis, and
has to do with "rational" signs only. The phenomena perceivable only by a 'scope or speculum,
the sounds elicited only by percussion and auscultation, are as truly symptoms as is a dilated pupil or
a wheezing respiration. Not thus, but because to
ascertain the seat of disease we have to bring in the
aid of morbid anatomy. This is the science of *lesions*,
while clinical medicine takes account of *maladies* —

which, in the words of Tessier, are "constituted by
an assemblage of symptoms and lesions undergoing
a definite evolution." The one speaks of hepatisa-
tion of the lung, the other of pneumonia; the one
of herpes, the other of shingles. Now the lesion —
save where, as in the last instance, it is on the
surface — is a thing inferred only, not perceived
or experienced; and hence is not strictly included
within the range of the knowledge of disease required by the homœopathic method, which — again
to quote Tessier — is one of "positive indications."
To many minds, accustomed to make physical diagnosis their chief aim as physicians, this is a very
unacceptable feature of our practice. But let us
look at the matter dispassionately. What do you
gain by inferring, from certain signs, that a given
group of symptoms means the presence of inflammation of the air-cells proper, as distinguished from the
bronchial mucous membrane or the pleura? Something, it may be, for prognosis: you know better
what the patient has to expect, and both he and you
feel more security from being able to follow the morbid process as it were with your mind's eye through
all its stages. In other cases, as where the digestive
organs are affected, a knowledge of the precise seat
of the malady aids you in general management: you
can order such food only to be taken as will give the
affected portion rest — farinaceous diet where the
stomach, animal where the duodenum is involved.
In neither instance, however, have you gained anything as regards positive treatment, especially if you

are going to conduct this by similar remedies. Your
medicine must indeed act on the same parts as those
affected by the disease, and in the same manner.
But, if it produce a like group of symptoms, the
inference is that it does so. As Hahnemann wrote
in the *Organon* (§ 148) — "A medicinal substance
which has the power and the tendency to produce
symptoms the most similar possible to the disease to
be cured, given in a suitable dose affects those very
parts and points in the organism hitherto suffering
from the natural disease." It is from the phenomena
that, in diagnosis, you infer the noumena : quite as
surely, in treatment, if drug and disease have the
same phenomena, it may be concluded that their
noumena are also identical. You are indeed in this
way more certain of your aim ; for your diagnosis
may be wrong, as the autopsy not uncommonly
proves, but your comparison of symptoms — if intelligent and painstaking — cannot err of the mark.
And further, — it must be remembered that our
object is to select, not a *simile* only, but the *simillimum* — the medicine whose action on the healthy
corresponds to the particular case in its individuality,
in the finer features and more minute ramifications
of the malady here presented. Identity of lesion is
insufficient for this : "we want" as Dr. Drysdale
has said "a pathological simile far more exact and
qualitatively like than that afforded by mere coarse
morbid anatomy, which is common to all cases alike."
We get this by fitting together the variety of phenomena manifested in disease and in drug-action, by

"covering" the one with the other. We may not be able to explain why certain symptoms are present in certain cases : but we must believe that each has its proximate cause, and that the combination of such causes constitutes the individual malady from which the patient is suffering, and to which our drug must be fitted.

For drug-therapeutics on the homœopathic principle, therefore, symptomatology may justly supersede diagnosis, as being in many cases surer and in all more thorough. It gives us this further advantage, that it often enables us to attack maladies in their forming stage, before they have developed such lesions as physical signs can manifest. The totality of symptoms is intended to be a *curative* indication, and if disease is to be cured it should be taken as early as possible, before such results have occurred as become the subjects of morbid anatomy *post mortem*, or even of pathology during life. In such early stages maladies are recognisable by rational signs alone, and mainly by symptoms of a subjective nature. This point has been forcibly made by Carroll Dunham, in his essay entitled "The Relation of Pathology to Therapeutics ;" and I would take the opportunity of commending the writings of this "beloved physician" (by no name less tender can those who knew him speak of him) to your most earnest attention. His lucid style is but an index to the clearness of his thought ; and in him Hahnemann finds an expositor who knows how to reconcile him to science and expound him in reason without sacrificing an iota of

his essential principles. In the essay I have mentioned he shews, that as physiology takes cognizance,
not of life, but of the results of life, so that with
which pathology is concerned is the result of the
abnormal and perverted life which we call disease.
The products of disease pathology sees, hears, or
infers : it knows nothing of disease itself. Hence,
to base therapeutics upon pathology alone is to make
the former merely palliative — a pumping out a leak-
ing ship instead of stopping the leak. It may be
said that we do not know where the leak — the primary disturbance — is, and that if we knew we could
not reach it to stop it. But by the proving of medicines we obtain agents which shew their power to
cause similar inundations, and therefore, presumably,
similar breaches, which — upon the principle *similia
similibus* — it is the hypothesis that they can repair.
If, then, the comparison between the results of disease and of drug-influence be thoroughly and accu-
rately made, the parallelism of action must reach also
to that which originates either. "And here" Dr.
Dunham writes "I cannot refrain from rendering
homage to the wonderful prevision of genius by
which, in an age when pathology, as we understand
it, was unknown, Samuel Hahnemann anticipated all
that we have said, and all that the most advanced
thinkers of our day have taught, respecting the scope
and influence of Pathology in relation to Therapeutics. The symptoms of the urinary organs in connexion with the discharge of morbid urine would at
one time have been regarded as the proper subject

of treatment. But Pathology has now taught us
to trace these symptoms back to the kidneys, and beyond the kidneys to the blood, and beyond the blood
to the nutrition and the destruction of all the organised tissues. As Dr. Carpenter remarks — 'When,
for example, the urine presents a particular sediment,
our enquiries are directed not so much to the sediment itself, as to the constitutional state which
causes an undue amount of the substance in question to be carried off by the urinary excretion, or
which prevents it from being (as usual) dissolved in
the fluid.' To confine the attention, therefore, in
prescribing for a given case, to the immediate organ
the perversion of whose functions is most obviously
pointed out by the prominent symptoms, is to disregard the clearest indications of Pathology. We must
analyse these obvious symptoms, and must include
their remotest elements in our indications. Nay,
these remotest elements — the constitutional disturbances, for instance, of which Carpenter speaks — are
even more important indications for treatment than
the more obvious and objective symptoms. But
how can we analyse these more obvious symptoms,
and ascertain those 'constitutional disturbances' in
which they have their origin? In no other way than
by a study of the functions of the entire organism —
in what way and to what extent they are performed
in an abnormal manner. And this brings us at once
to that rule on which Hahnemann so strongly in-
sisted, that the entire organism of the patient should
be examined in every possible way, and that the

'totality of symptoms' should be made the basis of
the prescription ; nay, that the constitutional, general symptoms are often more conclusive as to the
proper treatment than the more obvious local symptoms. The grand old master reached at a single
bound the same conclusions to which the labors of
a half century of able pathologists have at last, with
infinite research, brought the medical profession."

All this time we have been dealing with general
principles ; but let us look at special forms of disease,
and see whether or no the Hahnemannian mode of
regarding them is sufficient for their treatment.

I. The *fevers* constitute a group which plays a large
part in daily practice. They are maladies in which
morbid increase of temperature exists prior or out of
proportion to any local inflammation which may be
present. The theory of this state is still a moot
one. According to some pathologists it depends upon
excessive heat-production ; according to others, upon
deficient heat-radiation ; while yet another class (with
whom I venture to think the truth resides) believe
that both factors operate in the process. But whatever be the genesis of fever, it remains a positive
fact, a clinical entity, with which we have to deal.
Upon the homœopathic principle, we have to treat it
with drugs capable of producing fever. How they
do so, we may not know ; but our ignorance of the
process matters little if we are sure about the result.
"An infinitesimal quantity of atropia —a mere atom,"
— writes Dr. Harley, "as soon as it enters the
blood, originates an action which is closely allied to,

if it be not identical with, that which induces the circulatory and nervous phenomena accompanying enteric or typhus fever." This is sufficient ; and as soon
as we learnt it to be a fact, from Hahnemann's provings of Belladonna (made, I may add, before Dr. Harley was heard of), yet minuter quantities of Atropia
(in the form of the juice of its mother-plant) became
in our hands trusted remedies for these very fevers.
Again (for here there is no question of lesions) the
classification of fevers of which we have already
spoken, so necessary for science and so valuable for
general purposes, has but the smallest influence upon
drug-selection. The old divisions of synocha, synochus and typhus (the last with its "nervosus" and
"putridus"), worthless as they are from a scientific
point of view, are much more useful for our practice
than those of typhus, typhoid, relapsing and ephemeral. They denote the *kind* of fever with which we
have to do, its quality and mode of life ; and to us it
is all-important that our drugs, next to being really
febrigenic, should correspond in their action to the
kind of fever present. They can hardly set up a
whole typhoid, in its complete evolution ; but the
febrile state they develope is certainly either a synocha, a synochus, a typhus nervosus versatilis or stupidus, or a typhus putridus ; and if we find these
states existing, in the essential fevers, the exanthemata, or elsewhere, in them we shall have our remedial means.

2. After fevers, the most important group of diseases consists of the *inflammations*. To the pathology

of this morbid process many pages are devoted at the commencement of every treatise on medicine or surgery. Whether, after all that has been said, we know much about it in its essence, may well be doubted; but, even if we do, of what avail is our knowledge for treatment — at any rate for medicinal treatment? The old phenomenal signs, *dolor, calor, rubor, turgor*, still for all practical purposes constitute inflammation, when externally manifested; and when it is internal, and so invisible, the facts which lead us to infer its presence and seat are no less of the symptomatic order, as I have already argued. To treat inflammation homœopathically, it is only necessary to find a drug capable of setting it up, at the same spot and in the same manner, as evidenced by the symptoms.

3. The *neuroses*, of which I would in the third place speak, are still — as Liebermeister says — symptomatic groups. Their unity is one neither of cause nor of (known) lesion : it is clinical only. It is of much interest to know what is the seat and process of the epileptic paroxysm; but our choice of anti-epileptic remedies must be determined mainly by the power they have of inducing similar paroxysms in the healthy subject, explain it or not as we can. In like manner is it with chorea and tetanus and hysteria : no conceivable knowledge we can gain as to their intimate nature would make us better able to fit homœopathic remedies to them than we should be if we possessed their symptomatic analogues in drugs.

It thus appears that, of the three elements we have seen to exist in all knowledge — phenomena, laws, and causes, it is the first which, for positive therapeutic action, chiefly concern us in disease. Not that the other two are worthless to us, even for this end. Our laws here are classifications — the recognition in morbid states of genera, species and varieties analogous to those of animated nature. These enable us to form groups of remedies associated with them, instead of having to wander through the whole Materia Medica for each prescription: they also give a continuity to medicinal treatment, without which the *usus in morbis* were of no avail. Hahnemann led the way here, by constantly insisting on the existence of fixed and definite types of disease, to which standing remedies should be applied ; and by giving us his group of "antipsorics." I fear, however, that he must be considered as having rejected all enquiry into causes — I mean proximate causes, the noumena of the phenomena — in this sphere. In so doing we need not follow him. His ground for taking symptoms as the element of parallelism between disease and drug-action was that they only were surely known. In his day this was true, and his selection of them was most prudent. But to maintain that they only were knowable was unwarrantably to bar the advance of science. His stricter followers have acted on the *dictum*, and have looked askance on the positive pathology of the present day, with its physical diagnosis and *post-mortem* confirmations. They are

always a decade or more behindhand in their recognition of such distinctions as those between typhus
and typhoid, between chancre and chancroid, and in
their use of such means as auscultation and ther-
mometry. Now this is altogether wrong. An inference from symptoms, if sure, is as good a basis for
treatment as symptoms themselves. This sureness
is assumed in the prognosis given and the general
management instituted : why should it not be also
for purposes of drug-selection ? By proceeding upon
it we secure another route to the *simile* we desider-
ate. We use symptoms to reach it, because they
are its most certain expression ; but, if it can be
otherwise attained, the alternative access may often
be useful. Morbid lesions sometimes occur almost,
if not quite, without symptoms, as for instance
caries of the vertebræ, and senile pneumonia. To
attempt to "cover" these from the results of the
proving of drugs would be futile. But toxicology
and experiments on animals here come to our aid,
and give us in Phosphorus a substance capable of
inflaming alike the cancellous structure of bone and
the pulmonary air-cells ; so that with it we can combat these diseases, however latent and expression-
less they may be. There is indeed something
fascinating about similarities of this kind ; and an
eminent English homœopathist, Dr. Sharp, has proposed (following in the footsteps of Paracelsus and
Rademacher) to make seat of action instead of
symptoms the basis of our method, which accordingly he would call "organopathy." That remedies

so led to may prove effectual is undoubted : we have a good example of them in the Ceanothus Americanus, which, though never proved on the healthy, and only known to "act upon" the spleen, has been found strikingly effective in enlargements and other disorders of this organ. But we should never, if possible, rest content with identity of seat between disease and drug : we should aim also at making their kind of action the same, and this can only be done by securing similarity in their symptoms. In this way we elevate the *simile* to a *simillimum*, and proportionately enhance its energy in cure.

We thus come back to the phenomena as our mainstay in practice : for therapeutic purposes, the totality of symptoms constitutes the disease. As a result of this view, the examination of patients by the homœopathic prescriber is far more minute than that ordinarily practised. He can hardly, indeed, inspect and explore for himself more thoroughly than does the well-trained practitioner of to-day ; but he listens to and questions the sick person with greater patience and more painstaking completeness. His object is not merely to find out to what recognised malady his patient's troubles are to be ascribed : for such end but few symptoms are necessary, and the rest can be left. He has to get at their totality, that he may cover them with a medicine capable of producing them on the healthy subject ; and in pursuit of this aim he must not account any detail superfluous. It has been objected that we should come off badly upon such a method with Mrs.

Nickleby for a patient. But happily Mrs. Nicklebys are exceptions ; and when we do meet them common sense must deal with them accordingly. Of course, proportion must be observed; and any thing we *know* to be merely incidental may be omitted. Our colours must be mixed, like Opie's, " with brains, Sir." But if we only *think* a detail unimportant, our wisdom will be to give the patient the benefit of the doubt, and insert it in our picture.

In this way we are led to that greater regard for subjective symptoms in favour of which Dr. Russell Reynolds so eloquently pleaded before the British Medical Association in 1874.

"Is it not coming to this " he protested "that but little atten-tion is often paid to the accounts which patients give of them-selves, their ideas, emotions, feelings, and physical sensations? These are things which we cannot weigh in our most guarded balances ; measure by our finest scales ; split up by our cruci-bles ; or describe in any terms save those which are peculiar to themselves, and which we cannot decompose. These symptoms are often disregarded and set aside ; and the patient, whose story of disease is made of them, is thought fanciful, hypochondriacal, hysterical, nervous, or unreal ; because, forsooth, we have physi-cally examined thorax, abdomen, limbs, and excretions, and have found in them nothing wrong; because we have looked at the retinæ, examined the limbs electrically, traced on paper the beatings of the pulse, weighed the patient and not found him wanting. Still he is miserable, in spite of placebo and assurance that there is nothing organically wrong! There may be in him a consciousness of a deep unrest ; or of a failing power, which he feels, but which we cannot see ; or of a something worse than pain, a sense of impending evil, that he is conscious of in brain or heart ; a want of the feeling of intellectual grasp, which

he may call failure of memory, but which memory — when we
test it — seems free from fault; a want of the sense of capacity
for physical exertion, which seems, when we see him walk or
run, to be a mere delusive notion, for he can do either well or
easily to our eyes and those of others; and so he is called nervous, and told to do this or that, and disregard these warnings
which come to him from the very centre of his life. And let
me ask whether or no it has not again and again happened in
the course of such a history as that which I have only faintly
sketched, that some terrible catastrophe has occurred? Do we
not see minds gradually breaking down while we say there is
no organic change in the brain? hearts suddenly ceasing to do
their work, when after careful auscultation we have said there
was nought to fear? Suicide or sudden death sometimes dis-
turbs the calm surface of our scientific prognosis of no evil:
we may be startled, and may then see all that we ought to have
seen before. But when the ripples that such unforeseen events
have occasioned on that smooth surface have subsided, we go
on as we have already done, and still pay but little attention to
what the patient feels, and delight ourselves in the precision of
our knowledge with regard to physical conditions of which he
may know nothing and may care still less. No one can appreci-
ate more highly than I do the value of precise observation, but
I do not believe that minute, delicate, and precise observation is
limited to a class of facts which can be counted, measured, or
weighed. No one can see more distinctly than I do the wrong
conclusions at which a physician may arrive by accepting as
true the interpretations which fanciful patients may offer of their
symptoms; but I am sure, that if we pay no heed to these mistaken notions of a suffering man, we lose our clue to the com-
prehension of the real nature of his malady. Morbid sensations
and wrong notions are integral parts of the disease we have to
study as a whole, and we are bound to interpret their value for
ourselves; but we can ill afford to set them aside, when we are
as yet but in the dawn of scientific pathology, and are endeavouring to clear away the obstacles that hide the truths we hope
hereafter to see more clearly about the mystery of disordered

life. The value of such symptoms may be slight in some kinds of disease, when compared with that of those phenomena which may be directly observed; but we are bound to remember that there are many affections of which they furnish the earliest indication, and there are not a few of which they are throughout the only signs."

In the light of this, which is but one among the many advantages of Hahnemann's mode of observing disease, I think we may make claim for it as being, not only the one safe thing for his own time, but also a mode of procedure most important in itself, and never to be left behind. It needs especially to be emphasised at the present day. It is with us as before the Reformation, when the Bible was used by the Church only as a rule of faith — a source whence were to be inferred the doctrines and practices obligatory on her children. What Luther and his fellows did was — as Dr. Robertson Smith has well shewn — to recover the Book itself, in the totality of its thoughts and words, as a means of grace to each individual soul. The fruitful results thus achieved in the spiritual sphere will be paralleled in the medical as the clinical study of disease is allowed its due preponderance, and is made the direct road to therapeutics. Of this reformation Hahnemann was the preacher in his day ; and his voice must ever be echoed by his disciples when they see the profession straying into the alluring, but less practical, by-paths of pathological speculation.

In support of thus acting, they can now cite the words of another acknowledged leader in English medicine, Sir Andrew Clark. In his Presidential Address at the Clinical Society of London in 1883, this distinguished physician said :[1] —

"Another great work of our Society has been, and continues to be, the gradual unfolding of the exact relations which morbid anatomy and, incidentally, experimental pathology should hold to clinical medicine. These two chief servants of our art, excited and carried away by their marvellous successes, and assuming a joint sovereignty over our art, look down with condescending superiority upon clinical medicine, ridicule her claims to supremacy, scoff at her empirical distinctions, reproach her with being unscientific, and strive to torture her into a slavish subjection to their theories. But the true relation is not this ; it is, indeed, the converse of it. For the structural change is not disease, it is not co-extensive with disease ; and even in those cases where the alliance appears the closest, the statical or anatomical alteration is but one of other effects of physiological forces, which, acting under unphysiological conditions, constitute by this new departure the essential and true disease. For *disease in its primary condition and intimate nature is in strict language dynamic ;* it precedes, underlies, evolves, determines, embraces, transcends, and rules the anatomical state. It may consist of mere changes in the relations of parts, of re-arrangements of atomic groupings, of recurring cycles of vicious chemical substitutions and exchanges, of new conditions in the evolution and distribution of nerve force, and any or all of them may be invisible to the eye, inseparable from life, and undiscernible in death. Undoubtedly the appearance of a structural alteration in the course of disease introduces a new order of events, sets in action new combinations of forces, and creates disturbances which must be reckoned with, even as mechanical accidents of the pathological processes. But *always behind the statical lies the dynamic condition ;* underneath the structural forms are the active changes which give them birth, and stretching far beyond

[1] *Lancet,* Feb. 3, 1883.

the limits of pathological anatomy, and pervaded by the actions
and interactions of multitudinous forces, there is a region teem-
ing with manifold forms of disease unconnected with structural
change and demanding the investigation which it would abundantly reward. It is in this mysterious and fertile region of
dynamic pathogenesis that we come face to face with the primitive manifestations of disease, and learn how much knowledge
from various sources is needed to understand it aright; it is
here that we see how, without help from physics, chemistry, and
biology, collecting, converging, and meeting in a common light,
no single problem in disease can be completely solved; it is
here that we are made to comprehend how the nature of a pathological product cannot be determined by its structural character, but by the life-history of the processes of which it is only
a partial expression; it is here that we observe how, in therapeutic experiments, *the laws of the race are conditioned and
even traversed by the laws of the individual;* and it is here
that we discover how clinical medicine is to become a science,
and how she is already, beyond question, at once the mother
and the mistress of all the medical arts."

IV.

THE KNOWLEDGE OF MEDICINES.

AT our last meeting we spoke of the knowledge of disease. We saw that the phenomena we call "clinical" — the symptoms of maladies, subjective and objective, rational and physical, in their connexion, conditions, and order of evolution — form the most important object of our study. They do more than enable nosology to classify their sum and pathology to diagnose their seat : they directly avail, under the guidance of the method of Hahnemann, for the choice of their remedies. Nosology aids in this, by grouping drugs around definite morbid species, and pathology by utilising their local affinities ; but both need completing by symptomatology to determine finally the one medicine which shall be the *simillimum* of the disorder we have to treat. We heard some of the ripest medical thinkers of our time bearing witness indirectly to the validity of this mode of procedure, recognising the dynamic origin of disease, the importance of subjective symptoms as indicating its beginnings, and the necessity of taking all symptoms into account if we are to arrive at a true con-

ception of a case. The inference is that to the clinical study of disease, as made possible by hospital and
dispensary, you should devote your chief attention.
Learn, indeed, all that pathology, which is the science
of disease, can tell you about it in its various forms ;
but use the light of such knowledge, not so much to
gaze upon in scientific interest, as to illumine your
perception of the actual features of that with which
you have to do.

Our subject to-day is the knowledge of medicines,
which are the tools of the healing art, as disease is
the material on which it works. What are medicines ?
I do not know that any better definition of them can
be given than that which was put forth by Hahnemann in 1805, in the preface to his *Fragmenta de
viribus medicamentorum positivis:* — " Quæ corpus
meré nutriunt, *Alimenta*, quæ vero sanum hominis
statum (vel parvâ quantitate ingestâ) in ægrotum —
ideoque et ægrotum in sanum — mutare valent, *Medi-
camenta* appellantur." My only difference with him
would be that I should place the corollary foremost,
and define a medicine as a substance which has the
power of changing sickness into health, and therefore — on the principle *nil prodest quod non læditur
idem* — of altering health to sickness.

Now on what ground is any substance to be reckoned a medicine ? and how is it to be ascertained
what are the morbid conditions and processes it can
favourably modify ? There are but two ways by which
to arrive at such conclusions, the empirical and the
rational.

1. Many, perhaps most, of the ordinary remedial uses of drugs have been stumbled upon by chance. It has generally been "the common man" (as Hahnemann calls him), sometimes even the still lower brute, that has discovered them ; and the professional healer has taken the hint and adopted the practice. After this manner has been gained bark as a remedy for ague, burnt sponge for goitre, arnica for the effects of falls and strains, graphites for tetters, sulphur for the itch. Nor less empirically, though among the practitioners of medicine, has arisen the use of mercury and iodide of potassium in syphilis, of bismuth in gastralgia, of arsenic in psoriasis. Theories of the *modus operandi* of such remedies have often been subsequently framed; but it is certain that their original adoption grew out of no such theories, but was an accidental discovery.

Now it would be the height of unwisdom to neglect information from this source. A remedy is a remedy, however come at, and whether conforming or not to any laws of action we may suppose to prevail. Experience is the test even of medicines rationally ascertained to be such : it is but beginning the process a little lower down when experience itself discovers them. But on the other hand it is obvious that the empirical method is a very uncertain one, and affords no guarantee of further additions to our remedial wealth. Indeed it is no method at all, but mere guess-work and chance picking-up. It is only hopelessness as to rational therapeutics which can lead writers like Wilks and Druitt to make empiricism a matter for satisfaction and a standard of advance.

2. There are certain pseudo-rational modes of discovering remedies which have brought undeserved slight on those truly bearing the name. Such are the doctrine of "signatures" and much of the iatro-mechanical and iatro-chemical theory of former and later times. When a real medicine has been gained by these means — as chelidonium in disorders of the liver and euphrasia in those of the eye, as iron in anæmia and muriatic acid for low fevers — it has been by coincidence, not from induction : the result is practically empirical. The truly rational method is that which infers the place and power of a drug in disease from its behaviour in health. Every such substance, on being introduced into the animal organism, causes certain disturbances, certain changes. Each has its proper series of effects : each selects certain organs and tissues, or certain tracts and regions, of the body, and there sets up phenomena of a definite kind. This is the only source of information which is surely and indefinitely fruitful. If from observing the pathogenetic effects of a substance we can conclude (subject to the teachings of experience) as to its therapeutic virtues, we have but to experiment with fresh poisons in order to gain as many additional remedies.

It is (as you know) the glory of Hahnemann that he perceived the supreme value of this mode of discovering medicines, that he earnestly preached and diligently practised it. By so doing, and by discovering — in *similia similibus* — the link between pathogenetics and therapeutics in the latter's highest form, he opened the way on which we are now advancing, and made possible every triumph we can obtain. I am not relating to you the history, or vindicating to you the truth, of Homœopathy; I assume that you know the one and recognise the other, and therefore I assume your concurrence with me in the proposition that our knowledge of drugs must be largely concerned with their physiological action. Their effects on the healthy body form the parallel to the phenomena of disease: here, as there, we must know these phenomena themselves, and what we can of their laws and causes.

And here again the phenomena, in their totality, are those which claim our most earnest attention. The clinical study of disease must have its complement in the acquaintance we make with medicines. The law of similars teaches us this, for to obey it we must have wholes to compare with wholes. In the ordinary practice men aim at knowing what drugs can do, that in disease they may induce such effects with them, when they judge it desirable. In old days, accordingly, they cared only to learn whether a given one could purge, or puke, or sweat, that they might class it as cathartic, emetic, or sudorific; and now they correspondingly limit their investigations to the question whether it is an excitant or depressant of certain nerve-tracts. For such purposes dumb creatures suffice; and hecatombs of these unfortunates are now annually sacrificed in enquiries as to drug-action. The differences of result, and therefore of opinion, are endless; and the gain is proportionately small. We, on the other hand, have wanted the whole picture of the effects of drugs for comparison with the phenomena of disease, and have gone to work accordingly. As it is human disease to which we need *similia*, and as this is largely made up of subjective symptoms, it is on the human subject that we experiment ; and we faithfully record the whole series of morbid changes which occur after the ingestion of a drug. We test the effect of single full doses, to get analogues of acute disease ; and of long-continued small ones, that chronic maladies may find their antitypes. Thus our pathogenetic knowledge, when truly obtained and registered, is like a picture-gallery, in which the discerning eye may perceive the lineaments of all morbid conditions known or likely to occur. Our provings minister to medicine as an art : they are synthetic and sensuous, full of colour and detail. Those of the other camp are rather analytic, appealing to the reason ; and are available only so far as morbid processes are scientifically understood. The record of the one recalls the graphic pictures of Hippocrates and Sydenham and Watson, to whose ever-fresh lineaments the mind returns with pleasure, wearied with the merely intellectual refinements of modern nosography. The work of the physiological laboratory goes hand in hand with that of the dead-house : the Hahnemannic pharmacology and pathology alike move in the region of life.

This is the first truth about the matter. To read a good proving in detail — like the Austrian of

Aconite — is like going through a series of clinical cases illustrative of the varieties of a particular disease. But this comparison of itself suggests that a further step is necessary. In recording such cases, a physician would do so by way of basis for a discussion of the disease in question, of its causes, its nature, and its treatment. For ourselves too we have seen that the clinical aspect is not the only one in which we should regard the ills to which flesh is heir. From pathology — the science of disease — its phenomena are always illuminated, and sometimes even rendered transparent so that through them we can see the noumena. Pharmacology should seek a stand-point no less advanced. Provings correspond with our studies at the bedside or in the prescribing room; but as to interpret these we go to the dead-house, so to the study of provings we must add — when possible — that of poisonings and of experiments on animals, that the lesions wrought by drugs may be positively ascertained. But yet our work is not done. We must use these facts also as materials for inductive generalisation; we must seek to connect, classify, and interpret them, to ascertain their laws, to trace them to their causes. In proportion as we do so, we make our pharmacology a worthy mate for the pathology which is growing into maturity beside it. In neither do we content ourselves with generalisations alone : the clinical history of diseases and the detailed provings of drugs must ever form the basis of our knowledge, be the superstructure what it may. But while

(to employ another figure) these constitute our text,
we should read it with the help of a commentary
which may illuminate it by the best available lights.
There are some who think they are best following
Hahnemann by shutting their eyes and ears to all
that has been learned since his time ; by recognising
nothing in disease but the patient's sensations and
obvious appearances, and nothing in drug-action but
a scattered heap of symptoms of like kind. We
should not go to the other extreme, and ignore any
aid which may thus be gained in practice. But we
should regard the human body, whether idiopathi-
cally or medicinally disordered, as one of whose
order we are not wholly ignorant — as a sphere in
which we are to some degree at home, and where
we may speak and act as no mere strangers. In
studying the Materia Medica we are to be more
than symptom-memorisers, in applying it more than
symptom-coverers. We are *cleri* and not *laici* here,
and we fall short of our vantage-ground if we work
mechanically only.

Let me illustrate. Dr. Jousset, in his recently-
published *Traité de la Matière Médicale*, describes
the action of Digitalis on the heart. He tells us that
in the form of poisoning he calls *foudroyante*, the
pulse is small, uncountable, sometimes completely
absent, and the heart-beats precipitate and hardly
perceptible, with irregularity as recovery ensues;
that in the less severe variety he designates *progressive* the pulse is at first strong and hurried without
irregularity, beating 120 to 140 in a minute; and

that when the effect is less pernicious, this rapidity is succeeded by slowness, which from small doses occurs at once. Again, in summing-up this part of the drug's action, he states that strong doses "paralyse the heart and the arteries after hav'ng excited them," while still larger doses paralyse from the first ; and that, of feeble doses, "strong and retarded cardiac impulse is the primitive, feeble and accelerated the secondary, effect." All this is true and useful enough, but what does it mean? The heart is a hollow muscle, contracting rhythmically under the influence of the ganglia embedded in its substance, and regulated by the opposing influences of the pneumogastric and sympathetic fibres coming to it from the central nervous system. What is meant by its being "paralysed"? Is it the cardiac muscle itself that is incapable of responding to the nervous impulses, or the nervous centres which have no power to send forth their commands? Is the alteration of the heart's pulsation due to inhibitory or accelerating influences transmitted to it from above, or to some change in the organ itself? These are questions of no mere speculative interest : on their decision depends our view of the use of the drug as a remedy. If it paralyses the cardiac nerves only, it cannot strengthen and tone up a dilated ventricle by its homœopathic action. If it retards the heart otherwise than through the vagi, slow pulse is no indication for it when induced by their inhibitory influence. Traube has actually ascertained that they are the channels through which Digitalis re-

tards the heart, and Claude Bernard found the drug
to be a direct muscle-poison. That the student
should know this gives him a precious clue through
the mazes of the phenomena displayed, and shews
him where the drug is truly homœopathic, i.e. where
the pulse is simply slow, or where the cardiac muscle
is of feeble vitality.

The physiological effects of drugs, with their interpretation, form thus the chief material of that
knowledge of medicines about which we are enquir-
ing to-day. But there is another and only less important field to be worked : I refer to the *usus in
morbis*, the therapeutic experience gained with the
substances we employ in practice. I have already
urged that a medicine — *medicamentum* — is one
which has shewn its power *medicare*, to heal : it
would be such did we know nothing of its power to
hurt. And conversely, though the pathogenetic
effects of a drug, when ascertained, are our most
trustworthy source for eliciting its curative powers,
they are not necessarily conterminous with these.
They may fall short of them, through inadequacy of
provings and absence of poisonings ; or they may
outrun them, from the multitude of trivial sensations the drug may elicit without definite character
or localisation. Of the former alternative we have
an instance in Hamamelis : how slight is its pathogenesis compared with the frequent use we make of
it ! For the latter we have examples ready to hand
in Ptelea trifoliata, in Fagopyrum esculentum, and
in one of the few provings we in England have

achieved — that of Cotyledon umbilicus. These drugs
have been freely and fully submitted to experiment : their pathogeneses in Allen's *Encyclopædia*
contain 965, 836, and 253 symptoms respectively :
but who uses them — at any rate the two last — in
practice ? The therapeutic virtues of a drug must
therefore be studied on their own merits ; and, when
they are of very definite character or on any thing
like an extensive scale, had best (I think) be taken
first.   The pathogenetic effects can follow ; and these
will throw light on the clinical results which have
been obtained, will give them precision and not
uncommonly expansion.

I need not say that in so doing you will avoid the
vicious practice of the old-school writers on Materia
Medica, who as a rule carefully abstain from connecting the physiological and therapeutic actions of the
drugs they treat of.   Of old, I imagine, they did this
from indifference and scepticism (which last Stillé
has plainly avowed) : but now the utmost charity
cannot acquit writers like Ringer, Phillips and Bartholow of deliberately adopting this course that the
frequent testimony their practice bears to homœopathy may not appear.   You, of course, will hail such
testimony rather than shut your eyes to it.   But
let me urge you not in your turn to be lacking in
candour when the evidence points the other way.
This caution is not needless.   Here is a patient in
the agonies of angina pectoris.   His heart is as it
were compressed, his breathing almost impossible :
his face is deadly pale, his surface cold, his pulse

small and contracted. Your old-school colleague
steps forward, and, taught by Dr. Lauder Brunton,
applies to his nostrils a few drops of the nitrite of
amyl. In less than a minute his face begins to flush,
he warms up, he breathes freely, and the intolerable
breast-pang is gone. This is beautiful practice : but
is it homœopathy? Nay : for let a healthy man
inspire the same substance, and the effect will be,
not the pallor and coldness and constriction, but the
flush — the dilatation of the imprisoned arteries —
which delivered the sufferer. In all common sense
and justice, therefore, this action ought to be ascribed to the second of those three *modi operandi*
of medicines described by Hahnemann : it is enan-
tiopathic, antipathic, affording all the speedy palli-
ation characteristic of such remedies while open to
all their disqualifications and reproaches. I have
nevertheless been grieved to see more than one
communication to our journals which, on the strength
of some incidental phenomena in disease and drug,
have claimed the action of amyl in angina for the
law of similars. My colleague Dr. E. M. Hale would
go farther still ; and by his theory of "secondary
homœopathicity" would catch every application of
the physiological action of drugs in the Hahnemannic net. I admire his enthusiasm ; but I deplore
the effects of such teaching. It leads, I fear, to a
great deal of very eclectic practice ; and it lays us
open to just retaliation on the part of our *confrères* of
the other camp. We say to them, — you are taking
our similar remedies, small dose and all, and refusing

to acknowledge the law under which they act, using them empirically, or explaining away their apparent homœopathicity.   They will say in return, — you are taking our contrary remedies, full dose and all, under a plea which to us at least is transparently futile : and I do not see how we can repel the allegation.

You will not hesitate, therefore, to give credit to anti-pathy when you meet with it ; nor will you ignore any other curative applications of drugs because they seem to lie outside the homœopathic method.   Whether you should employ them must depend upon other considerations, which you will learn from your teachers : just now, however, this is not a practical question for you.   Putting these aside, you will find in your therapeutic studies of medicines abundant examples of true homœopathic action, and will enrich your knowledge of them accordingly.   You will also learn here another thing which pathogenesy alone could not teach you : I refer to what are known as the "characteristic symptoms" of our remedies.   A great deal too much has indeed been made of these features : conditions have been dealt with independent of the things conditioned ; adjectives have been treated as more important than their substantives ; and, so long as disease and drug strike the same "key-note," it has been reckoned indifferent whether or not they play the same tune, — with what results in discord it is not difficult to imagine.   But, on the other hand, given your substantive, an adjective peculiarly belonging to it is often of value in suggesting it and — in case of need

— distinguishing it from others of its kind. Hahnemann initiated such characteristics when he recommended Aconite to be given in inflammations when "with thirst and rapid pulse, an anxious impatience, an unappeasable restlessness and an agonised tossing about are conjoined." Experience has abundantly confirmed the indication : Dr. Dunham has shewn its coherence with the pathological condition to which the drug is suitable, and Dr. Guernsey has under its guidance extended the range of Aconite into regions which might otherwise have seemed foreign to it. The aggravation of Bryonia pains by motion and those of Rhus in rest ; the early morning waking of the Nux vomica patient, the evening exacerbation of Pulsatilla sufferings, the intolerableness of those of Chamomilla — these are other well-known and well-tried characteristics which we owe to the master. Later experience has added others quite as trustworthy — the aggravations of Rhus symptoms on change to wet weather, of Rhododendron symptoms at the approach of storms, of Lachesis symptoms after sleep ; the relief of the Coffea toothache by cold water held in the mouth, the nausea of Colchicum at the smell of food, the sensitiveness of the Hepar patient, and so forth. Dr. Claude has lately shewn us, by a series of well-selected cases, how important an indication is what he calls the rhythm of medicines, — as the tendency of Lycopodium attacks to supervene from 4 to 8 P.M. and of those of Belladonna to come on somewhat later. Dr. Hawkes (of Chicago) has in like manner illustrated the unfailing way in which a certain group of symptoms — vertex headaches, with heat there, burning in the soles of the feet in bed, hot flushes, and "gone, empty feeling" about an hour before the midday meal — characterises a patient in whose ailments Sulphur will be beneficial. I might continue such an enumeration *ad infinitum;* but the point I desire to make is that all — or nearly all — this knowledge is gained from the *usus in morbis*, and would not be learned from a study of pathogenesy alone. You must therefore cultivate this field assiduously if you would fill in the outline of your acquaintance with the medicines you are to use.

So far I have been speaking as if disease and drug-action constituted a precisely parallel series of phenomena. There is an important difference between them, however, which perhaps has already occurred to your minds. Disease is ever with us. Though it is not exactly producible at will, its originating germs, or its predisposing and exciting causes of a general character, are so widely diffused, that no one need go through his student days without having seen one or more cases of at least its ordinary forms. In exhorting you therefore to study disease clinically, with pathology as the lamp to illuminate the field, an intelligible and practicable task is being assigned. With drug-disease it is otherwise. This *is* to some extent producible at will, but the will is (naturally) lacking; and we can only study it in the records left us by those who have been — voluntarily or involuntarily — the subjects of medicinal action.

This is a serious drawback ; and the only compensation for it would be that the records in question were
as full and clear and life-like as possible. Suppose
the student were debarred from studying disease at
the bedside or in the dispensary. He would value the
more the teachings from the chair of the Theory
and Practice of Physic ; but he would feel that, after
all, what he got there introduced him to maladies
only in their typical, somewhat abstract and ideal,
forms. He would crave for lectures and books containing detailed narratives of cases, so selected as to
illustrate the various forms and varieties in which
disease is liable to occur. But suppose that, instead
of such material, a volume (or ten volumes) were
put into his hands in which, under the head of each
malady, there was given a list of all the symptoms
which had been observed in the several instances of
its occurrence, — these being divorced from their
connexion and sequence, and re-arranged under the
headings of the part or function of the body to which
they seemed to belong. Would he not reject with
loathing this stone offered him instead of bread ?
would he not clamour for the clinical records, fresh
from nature's mint, of which some cruel technicality
had gone out of its way to deprive him ?

You will have anticipated me in feeling that this
is the unhappy position in which you are actually
placed in respect of the Materia Medica. I know
your teacher of that subject too well to doubt that
he gives you, in the most instructive way, all the
information regarding it which is communicable in

lectures. But, after all, he has only introduced the medicines to you : he has done for their pathogenesy what the Professor of "Theory and Practice" does for disease. Where, then — as you cannot study drug-maladies in life — are the clinical cases, the detailed provings and poisonings of individuals? They mostly exist, — though Hahnemann is said to have been un-kind enough to destroy the day-books of his provers : they exist, — but scattered through interminable vol-umes, often sealed up in a foreign tongue, sometimes still only in manuscript. The writer of a monograph on any drug has the utmost difficulty in getting to-gether all the necessary materials, as a certain Club in this city found when they began the labours which resulted in their charming little volume on Gelsemium : [1] but what is the student of all drugs to do? He may make the most of such books as those of Hempel and Burt, or as that *Manual of Pharmacodynamics* of my own which the authorities of this school have honoured me by admitting among their text-books. But these, after all, occupy only the same ground as the lectures he hears, and must be read in connexion therewith. Both one and the other are *introductions:* but what is it to which you are introduced? Whether in Hahnemann himself, in Allen, or in Hering, it is that unhappy jumble of symptoms whose inadequacy to instruct has been already seen in the parallel instance of disease.

What, then, are you to do? As matters stand at

---

[1] *Gelsemium sempervirens.* A monograph by the Hughes Medical Club of Massachusetts. Boston : Otis Clapp and Son.

present I would advise you, for all *à priori* knowledge of medicines, to content yourselves with the general view given in your lectures and text-books, with the broad features of their physiological action eked out with the fullest possible acquaintance with what they have done and are likely to do in the way of therapeutics. Their minute symptomatology it is hopeless to memorise, but no less perilous to abridge: let it stand as it is, with such discrimination as you can supply, and use it *à posteriori* only. Whenever an unusual symptom or combination of symptoms meets you, whenever a case "hangs fire" under the ordinary remedies, hunt up the phenomena by means of a repertory, which is just an index to the Materia Medica. If you find them there, if their source (so far as known) is trustworthy, if the drug to which they are ascribed is otherwise suitable to the case, give it, nothing doubting. You will sometimes miss your mark, but quite as often you will hit it; and in this way you will turn to profit a symptomatology which otherwise is only a barren and trackless wilderness.

I believe it would be a great deliverance to the student, and also to the would-be convert, of homœopathy if it were distinctly understood that our symptomen-codices were not meant to be studied. The endeavour so to use them has turned back many an enquirer in disgust: as it is said to be ominous that our English Marriage Service begins with "Dearly beloved" and ends with "amazement," so is the schema reflected in the mind of its peruser,

who begins with "vertigo" and ends with "rage."
Hahnemann certainly intended his for no such purpose : it is demonstrable that the *à posteriori* use of
them was the only one he contemplated, and that it
was in view thereof that he published his provings
in this form. Let us hold his works in all honour.
Let us give them adequate translation into our own
tongue, as we in England have lately done with the
Materia Medica Pura, as I hope America will speedily do with the *Chronic Diseases*. But let us use
them aright ; not exhibiting them to outsiders as a
specimen of what Materia Medica should be, not
breaking our own or our pupils' hearts by vain attempts to learn from them what medicines can do.
For this purpose his prefaces and notes are often of
the utmost value, for he wrote them from a knowledge of the detailed provings. But the catalogues of
symptoms into which he cut up the latter he made for
reference, and for reference only should they be used.

Again let me illustrate. In 1847 Dr. Molin, of
Paris, published the following experiment made on
himself with Tartar emetic.

"Being in a good state of health, my pulse 64, I took at 8
A.M., fasting, five milligrammes of tartar emetic in water. This
dose was repeated for five days without perceptible effect. The
sixth day I felt nothing until about 4 P.M. The respiration
then appeared to me a little less free. Feeling no further effects,
and my appetite continuing good, I took about 9 P.M. a dose of
one centigramme. The night was passed in a restless manner,
and the sleep interrupted by a fatiguing heat; I felt necessitated
to drink several times, the respiration was slightly impeded; on
rising, general uneasiness, weariness similar to what follows a

febrile fit, the mouth clammy. At 8 A.M. I took one centigramme. No appetite: a simple soup for breakfast without relish. All the day I was in the same state. About 5 P.M. greater
uneasiness, especially about the epigastric region; nausea; desire to vomit but without result; respiration more impeded;
short dry cough, pretty frequent; great thirst; heat in the head;
white tongue; drinks appear always too sweet; clammy mouth;
two loose evacuations during the day; palpitation of the heart;
bruised feeling and general weariness, compelling me to go to
bed at eight o'clock. The ear applied to the chest gave evidence
of nothing abnormal, except that the respiration appeared much
too rough. At 9 P.M. I took 5 milligrammes. Agitated sleep,
difficult respiration, feeling of pressure on the chest during sleep.
At 5 A.M. I was awakened by a violent rigor, it lasted twenty
minutes, and was followed by heat; the pulse, which had been
little affected during the two previous days, increased to 78, was
full and strong; skin hot; face red; thirst urgent; heat in the
head; pretty strong palpitation of the heart; slight burning at
the stomach, — fulness, and inclination to vomit; respiration
very much impeded; feeling of pressure and constriction of the
chest; cough frequent, and a little moister; on auscultation, the
respiration appeared rougher than on the previous evening, and
deep inspiration was accompanied by slight pain under the left
nipple. Night very agitated; nightmare; disagreeable dreams.
I felt much the same in the morning as I had the previous day,
but deemed it advisable not to carry the experiment farther.
During the subsequent days the following symptoms occurred:
— the tenth day, no stool; towards evening, pulse 72; respiration somewhat less difficult; cough the same; hardly any pain
in the side; great thirst; a good deal of uneasiness; no inclination to vomit; night a little less restless. The eleventh day, a
little less roughness of breathing on auscultation; cessation of
the pain; pulse nearly normal; skin still hot; thirst less; uneasiness diminished; appetite in part returned; respiration still
obstructed; cough a little less; the night more tranquil. Twelfth
day, appetite; breathing nearly free; the cough continues;
scarcely any thirst; tranquil night. The symptoms continued

to diminish the subsequent days, so that by the eighteenth there remained no trace of indisposition, except slight cough, which persisted some time longer."

Now it is quite clear that in this proving of Tartar emetic Dr. Molin developed in himself an incipient pneumonia ; and the rest of the symptoms (save perhaps those of the gastro-intestinal organs) were *syndromata* of that affection. Read in their connexion and sequence, they are easily understood aright ; but imagine them cut up for the schema. Imagination is necessary here ; as — very fortunately — Dr. Allen overlooked Dr. Molin's experiments when compiling the pathogenesis of Antimonium tartaricum for the first volume of his *Encyclopædia,* and so has reserved them for the appendix in the tenth volume, where he gives them in detail as recorded. But the picture is not difficult to construct. The rigor, heat, full pulse, and thirst, would appear in the " Fever " section, and would seem to shew that Tartar emetic was capable of exciting primary pyrexia, and this indeed — seeing that there was fever also on the previous night — of an intermittent character. The agitated nights, with their troubled dreamings, would be classed under " Sleep," and would suggest that the drug had power directly to disturb this function of the brain ; while the palpitations would mislead similarly as to its action on the heart. The respiratory affection would itself be dismembered ; for cough is always, in our schemas, found under the head of " Larynx," while the rest of the symptoms would be referred to the chest.

Said I not truly then that our pathogeneses, in the Hahnemannian arrangement, are not intended for study? Their unity is deliberately broken up, that they may become an index rather than a text; and they must be used accordingly, for reference and not for consecutive reading. To a great extent, as we have seen, the sacrifice is made in vain; for the separate symptoms divorced from their context are often unintelligible and even delusive. All that is required would have been given by a short repertory affixed to each pathogenesis, or group of pathogeneses, shewing where — if anywhere — individual symptoms might be found. In this way the peculiar sensation of the sweet taste of drinks, experienced by Dr. Molin, might have been noted; and might have led to a minute adaptation of the drug. Hahnemann did something of the kind for the pathogeneses of his *Fragmenta de Viribus*, but he had already adopted for these the schema-form. It is too late for any alteration in what we have from him: but it is not too late to stop any presentation of new provings in this shape. I had hoped, until I saw the Transactions of the American Institute of Homœopathy for 1881–2, that no one would now inflict such a wrong upon us. It seems, however, that some of those who call themselves "Hahnemannian" feel bound to reproduce, with Chinese accuracy, the defects as well as the merits of their eponym. Their vicious procedure will call down its own punishment; for in the Materia Medica of the future such contributions, if admitted at all, will pretty certainly receive the comparative discrediting of smaller type.

Of this Materia Medica, on which the hearts and hopes of many of us are now set, I shall have to speak to you in the concluding lecture of this course. I shall then endeavour to shew you how you may have to your hand records of drug-action answering to the clinical aspect of disease, which you may study with equal interest and instruction. In the mean time, as I have said, — make the best use you can of your introductions, and of such detailed provings and poisonings as you can pick up from books or journals. But do one thing more. I have said that the student of disease, if put off with an artificial symptomatology, would *clamour* for the natural records of which he had been deprived. I want you also to clamour. If what I have said on the subject has commended itself to your minds : if on hearing such a narrative as that of Dr. Molin's you have thought *O si sic omnia !* let your voice be heard, and *sic omnia* will be. I am pleading for the constitution of an entire Materia Medica in that form ; and this largely in the interest of the students a thousand of whom annually fill the Homœopathic Colleges of America. If they will back my plea they will convert it into a demand ; and the supply will assuredly follow.

V.

PYREXIA AND THE ANTIPYRETICS.

WE have enquired what should be our knowledge of disease and our knowledge of medicines, and have arrived at certain definite conclusions thereupon. From what I know of my friends, the Professors of Practice of Medicine and of Materia Medica in this School, I have no doubt of being in agreement with them as to what you should learn in these departments, and as to how your knowledge should be acquired. I am no less sure that, under their guidance, you are becoming as thoroughly acquainted with disease and with drug-action as it is possible for students to be. I have nothing to add to what they teach you. I propose rather, on most of the remaining occasions of my addressing you, to occupy your attention with a series of topics intermediate between the spheres in which their instructions move. I propose to study certain *groups* of medicines — groups formed naturally out of relationships to morbid states or actions on the same parts of the body ; and to institute such comparisons between the several members of the groups as shall bring out their individualities, and thus ensure their accurate adaptation to the disorders they have to remedy. Fever and rheumatism are the morbid states I shall utilise for the purpose; while for action on the same part of the body I shall study the cerebral symptoms of a number of drugs in the light of the researches on cerebral localisation which have excited so much interest of late.

Our subject to-day, then, will be the group of drugs which, from the control they have shewn themselves capable of exercising over the febrile process, may be called antipyretics. Before approaching them, let us dwell for a time on the process itself, and ascertain what is known of its inner nature and of the manner of its occurrence.

That fever, pyrexia, consists essentially in increased temperature of the body appears from its name, which in both Greek and Latin points to combustion or glowing heat (πυρετός, from πῦρ, fire; *febris*, from *ferveo*, I glow). We shall see presently that something more than heat of body is required to constitute the clinical entity we call fever; but, though there may be hotter blood (as ascertained by the thermometer) without fever, there cannot be fever without hotter blood. Now this organism of ours is so nicely adjusted, both in its inner relations and in its re-actions with the environment, that the temperature of our bodies is at all times and under all circumstances almost uniformly the same. Whether we shiver at the poles or swelter within the tropics, whether we glow with exertion or feel the chilliness of sedentary and indoor life, the thermometer in our armpits tells the same tale, and marks 98.4° of Fahrenheit's scale, 37° of the centigrade, as our normal heat. There are, of course, oscillations about this fixed point, according to time of day, meals, exercise, and so forth; but they observe very narrow limits, rarely transcending two degrees (up and down) of the one scale or one of the other. This uniformity is secured by a due compensation between the production of heat and its loss. When production is stimulated — as by external warmth or physical exertion — the perspiring skin (with its twenty-eight miles of tubing) allows a freer radiation : when heat-formation, in coldness or quietude, is small, the dry surface and contracted cutaneous arterioles restrain the loss of what there is.

It would seem, therefore, that for fever to exist one or both of these compensating functions must be disordered. There must be either increased production of heat, or diminished loss, or the two must coincide. Both of the first two alternatives have had their advocates. As regards diminished loss, — restriction of heat-radiation may undoubtedly cause elevation of internal temperature, as has been ascertained experimentally during exposure of the skin to cold air or water; but the increase is too moderate to attain of itself the febrile height. I think, however, that it may be sufficient to set going, in vulnerable tissues, the excessive heat-production which the opposite theory requires; and I apprehend that this is the rationale of the simple fever which results from a chill. While in your country in the summer of 1876,

and sitting in the usual perspiration which the "heated
term" of that year developed, I went out on a piazza
to see a thunderstorm advancing over the country.
The strong blast of wind which so often heralds such
a storm swept down upon me, and closed my open
pores with irresistible force.   The next day I felt that
I had "caught cold," and in two days more coryza
had developed itself, and I had a temperature of 102°.
This fever was out of all proportion to the local
symptoms, which were quite moderate.   It had been
forming, I apprehend, ever since my chill; and its
starting-point was the check to heat-radiation which
then occurred.

I would thus explain "catarrhal fever" by diminished loss of heat as its primary cause; but I have
already said sufficient to shew that increased heat-
production must also be set up to constitute the fully
developed malady.   Without it true febrile tempera-
tures cannot be attained; and that it is present in
fever is shewn by the enhanced tissue-waste manifest
therein.   Excess of urea in the urine precedes rise
of temperature and outlasts its decline, and ranges
from $1\frac{1}{2}$ to 3 times the amount of its proportion in
health.   What does this mean — there being no
change in food to account for it — but increased
oxidation? and, if Wagner be right in saying that
"the essential sources of heat in the organism are
chemical processes, based upon oxidation," then any
excess in this action implies the production of an
abnormal quantity of caloric.   Nor would the case
be altered if the protoplasmic doctrine of life were

to lead us — as some think it does — to refer heat-production to the direct metabolic action of the living matter. Excess of tissue-waste must still imply increased metabolism of tissue, and with this must come increased evolution of heat.

To hyperoxidation, then (this conception being provisionally maintained), we must look as the main source of the undue heat present in the febrile state. In catarrhal fever, it is probably secondary to disorder of the apparatus which permits of heat-radiation ; and in fevers dependent upon local inflammations (including those of hectic type) the tissue-destruction may be limited to the seat of mischief. But in the essential, toxæmic fevers — in typhus, typhoid, variola, scarlatina, and their congeners — I take it that we have hyperoxidation of the blood, or the tissues, or both, as the primary factor of the morbid process. The virus of their contagion, when imparted to a susceptible subject, acts as a spark to the combustible elements, and the mischief slowly or quickly spreads. These fevers are thus more prolonged than those of catarrhal origin, — the latter subsiding on the supervention of perspiration, which liberates the retained heat and so removes a main factor of the whole trouble.

But there is yet another possible source of increase of temperature ; and that is the nervous system. I do not mean so much that portion of it which, from its influencing the calibre of the arteries through their muscular coats, is called *vaso-motor*. This, of course, is largely concerned in the heat-regulating

function of the surface, in which the blood-vessels
play so important a part. At one time their contraction, constituting the cold stage, and their subsequent dilatation, which forms the hot stage, of
fever, was supposed to be of the essence of the
process. But it is now known that the chill may be
altogether absent, so that the hot stage is no mere
re-active dilatation of the superficial vessels ; and, on
the other hand, when the chill does occur, it is found
to be an evidence that the temperature has already
risen. Its only causal influence can be some amount
of heat-retention when it is prolonged. Again, after
paralysis of the vaso-motor nervous system by destruction of the principal centre, no fever is induced
unless the animal be placed in a hot room. He is
unduly sensitive to his environment, but is not otherwise febrile : indeed, if the temperature be low, he
will die of cold. But experiment seems to shew that
there are heat-centres in the spinal cord independent
of the vaso-motor nerves, and that their injury is
capable of setting up a great increase of the bodily
heat. Sir Benjamin Brodie found that at the end of
forty-two hours after crushing the lower part of the
cervical enlargement of an animal's cord, the temperature (centigrade) was 43.9 — the norm being, as
you know, 37.0 ; and Billroth, Simon and Naunyn
have seen corresponding effects from injuries of this
kind in the human subject. Mr. Teale, of Leeds,
reported a remarkable case to the *Lancet* in 1875, in
which the temperature continued above 108° for some
weeks, sometimes rising to 122° and over. This, too,

was one in which injury to the spine had occurred. I have mentioned these facts, though I do not think they have much bearing on ordinary fever, — Mr. Teale's patient, for instance, having no very marked febrile symptoms; but they possibly account for the occasional supervention of that hyper-pyrexia which has frequently been noted of late, and which puts the patient in such peril. Even for this, however, we must have two factors — some pyrogenous matter in blood or tissues, *and* collapse of the heat-controlling nervous centres; for Mr. Teale's patient, in whom the latter element presumably existed, survived through weeks of a temperature which, e.g., in acute rheumatism would kill in twenty-four hours.

There are two other types of fever for which also we may look in this direction; viz., the hectic and the intermittent. Either is marked by a series of paroxysms made up of more or less chill, heat, and sweat, usually occurring in this sequence, and having a rapid rise and fall of the temperature of the blood as their basis, — the rise being here as elsewhere accompanied by increased excretion of urea. In hectic, we have a local infective process (as tuber-culisation of the lungs) where suppuration is going on; and any continuous pyrexia which is present may be explained as is that of the fever symptomatic of inflammation of any part. For periodical parox-ysms of the kind, however, we must invoke the agency of the nervous system; for we can hardly suppose, as some have done, that the pyrogenic

matter accumulates and discharges itself with such
rhythmical regularity as seen, for instance, in the
diurnal pyrexia of phthisis. The malarious fevers
may be similarly regarded. Here, too, the exciting
cause is probably of a substantive character, i.e., it
consists of an abundance of low forms of vegetable
life; but here too, I think, we must look to the
nervous system as the source of the paroxysms. It
may be that thus is explained the fact so often
noted, that it is easier, with the attenuated remedies
of homœopathy, to cure chronic intermittents than
acute ones. The latter have the cause still present
and in operation : the former consist rather in a morbid habitude of the nervous centres, which a strong
mental impression will often remove as effectually as
an appropriate drug.

From the nature of fever we may now pass to the
forms under which it is manifested. The classification generally accepted in the last century was that
an outline of which you see on the board.

Symptomatic . . { Inflammatory.
                  Hectic.

Idiopathic . . . { Intermittent.
                    Remittent.
                    Continued . . { Synocha.
                                    Synochus.
                                    Typhus . . { Nervosus. . { Versatilis.
                                                               Stupidus.
                                                 Putridus.

"Symptomatic" fever was that obviously dependent on some local inflammation; and, if continuous,
was known simply as "inflammatory," while, if it

occurred in a succession of daily paroxysms, it was called "hectic." "Idiopathic" fevers were those apparently of primary origin; and these too were divided according as their phenomena were "intermittent," "remittent," or "continued." Continued fevers were further subdivided on the basis of the character of their symptoms. If these were of the simple and sthenic kind familiar in inflammatory fever, the term "synocha" was used to designate the patient's illness. If of a somewhat lower type, "synochus" was substituted as their designation; leaving "typhus" for the well-marked "low fever," and adding "nervosus" or "putridus" as the stress of the disease seemed to fall on the nervous centres or on the blood. The "versatilis" and "stupidus" further qualifying the typhus nervosus need no explanation.

Such a classification is obviously unsuited for nosology, when once the essential nature of certain fevers, and their dependence upon definite miasms or contagions, is recognised. The distinction of symptomatic and idiopathic pyrexia still indeed holds good, and pyæmia and septicæmia find appropriate place as varieties of hectic. But intermittents and remittents are now classed together as malarious; while continued fevers are recognised as occurring under the four forms of ephemera, relapsing fever, typhus, and typhoid, to which some would add a "common continued fever" — the *fièvre synoque* of the French, the "gastric fever" of popular English speech. Thus we get the second schema presented to us.

Symptomatic ..	Inflammatory.	
	Hectic.	
	Pyæmia.	
	Septicæmia.	
Idiopathic ...	Malarious.	Ephemera.
		Gastric.
	Continued ...	Relapsing.
		Typhoid.
		Typhus.

Now, speaking generally, there is a tolerable coincidence between the apparent and the real types. Ephemeral fever is synochal in character; relapsing and gastric fevers would of old have been called synochus; while typhus and typhoid commonly present the characters of the typhus putridus and nervosus respectively. But, while this is so, we must not let the ancient distinctions be swallowed up in the modern, as though wholly obsolete. While the latter are all-important for prognosis of the course and probable termination of fevers, and for their general management, the former still hold good for therapeutic purposes. They are symptomatic, and therefore lend themselves with great appropriateness to a method of drug-selection like ours which uses symptoms as its materials. They also enable us to embrace such fevers as the catarrhal and rheumatic, and that accompanying the contagious exanthemata, which — though not finding place in the usual classifications — are no less genuine clinical facts. The same may be said of those recognised varieties of our common

continued fevers which are now referred to the
"typhoid" or "gastric" category. Trousseau gives
them as "mucous," "bilious," "inflammatory," "adynamic," "putrid," "ataxic" and "malignant." Our
own Trinks, to whom we owe a valuable study of
"abdominal typhus" (i.e. typhoid) in its drug-rela-
tions,[1] describes it as occurring under the forms
"simplex," "biliosus," "pituitosus," "putridus,"
"nervosus versatilis" and "nervosus stupidus." While
the essential fever thus manifesting itself may be
one and indivisible, the various forms under which it
appears are no less realities, and require a suitable
adjustment of our drug-remedies, as they do of those
of a more general kind.

Coming now to the treatment of fevers, it is first
of all necessary to recognise that pyrexia, as such, is
an evil state, fraught with untold injury to its subject. We know how twenty-four hours of it, in its
simple catarrhal form, will weaken a strong man; and
after the three weeks of it in typhoid Liebermeister
well describes the body of the convalescent as "emerging like a wreck from a storm; on the one hand,
having to throw overboard ruined parts and clear the
deck, on the other, to restore sails and rudder." It is
most desirable, therefore, to abate its severity and — if
possible — shorten its duration. In ordinary practice
this is aimed at by cold baths, or by drugs like quinine and salicylic acid. It is a rough, crude treatment, directed — as Dr. Cretin has justly pointed out
— against one element only in fever, viz.: the high

[1] See *Brit. Journ. of Hom.* vol. XXIX.

temperature ; but, nevertheless, it seems to be better than expectancy. Sassetzky, of St. Petersburg, has lately instituted a special enquiry into its merits, with very favourable results. He found that invariably the cold bath diminished the elimination of nitrogen, and that a similar but far less marked diminution occurred in the cases treated by quinine and the salicylate of soda. The quantity of urine was increased by each method, but to the largest extent by the bath. The assimilation of the solid and nitrogenous constituents of milk was improved under the bath treatment, as shewn by the very marked diminution in the fæcal elimination of nitrogen ; and the same fact was also noticed, to a less degree, under the other methods. The quantity of water ingested was diminished, and also the loss of water by the lungs and skin, except under the salicylate, when the cutaneous loss was increased. These are results which shew that, during the febrile state, the patient is in a far better condition under the antipyretic treatment than without it. It is for us to demonstrate that we can, by non-perturbative internal medication, do as well as or better than our old-school colleagues ; and spare patients the distress of cold bathing and the injury of the necessarily large doses of quinine and the salicylates. Comparative studies on a large scale are the only means of *proving* that we can do so ; but in the mean time we can study our available means, and see what are at least their positive virtues.

We turn accordingly to our antipyretic medicines, which we may reckon as fifteen in number, viz. : —

Acidum muriaticum.	Cinchona (and Quinine).
Acidum phosphoricum.	Crotalus (and Lachesis).
Aconite.	Gelsemium.
Agaricus.	Hyoscyamus.
Arsenicum.	Rhus.
Baptisia.	Stramonium.
Belladonna.	Veratrum viride.
Bryonia.	

They must be considered in their relation to the febrile process itself, and to the kinds and shapes in which it has come before us. And this we shall best do by taking certain representative medicines, and counting them as types, under whose headings the others may find their appropriate place. Such typical drugs are Aconite, Belladonna, Arsenic, and Bryonia.

I. The history of *Aconite* as an antipyretic is the history of homœopathy. What was known of it up to 1805 had led to its use in a few chronic affections only. In that year Hahnemann published his *Fragmenta de Viribus*, which contained — among others — a pathogenesis of this plant. It shewed its power of producing a series of paroxysms of alternate chill and heat, "repeated two, three, or four times before the whole effect ceased, which it did in from eight to sixteen hours." In 1811, republishing this pathogenesis (with additions) in the first volume of his *Materia Medica Pura*, he prefixed some remarks pointing to the above-named features of its action as promising usefulness from it in acute diseases. Of what kind these should be he had not then perceived. But somewhat later, while treating some inflammatory

disorders he was led to the employment of Aconite
from the similarity of some of the concomitant symptoms with some in its pathogenesis, and he found its
administration followed by a great diminution in the
frequency of the pulse, and a cessation of the febrile
state. He followed up this hint; and in 1822 was
able to proclaim it the one sufficient remedy for states
against which the whole antiphlogistic apparatus of
that day (and you know how heroic this was) was
considered necessary — the "pure inflammatory fevers." It has continued to be so reckoned ever since
in the school which he founded, and has of late won
no small acceptance in the ranks of those who (at
least outwardly) reject his method.

Remembering this history, we turn over one of the
latest treatises on Materia Medica and Therapeutics
to see what is said on the subject. Dr. Bartholow,
in 1877, writes : — " The monopoly by homœopathic
practitioners of the use of aconite has roused a preju-
dice against it, which has discouraged its employment.
Aconite is, however, an antagonist to the fever-process ; it is not applicable in accordance with the so-
called law of similars. It is used by these quacks
because it is a powerful agent which will produce
manifest effects in small doses, that may easily be disguised." One hardly knows whether most to smile
at the ignorance or sigh at the insolence of this
statement. That Hahnemann was the discoverer of
the antipyretic power of Aconite, and that he arrived
at it by working with the law of similars, is manifest
from the facts of the case : that he required for its

exercise no doses which needed disguising, however easy, appears from his recommendation (in 1822) of the 24th dilution as sufficient for the purpose. That Dr. Bartholow's own knowledge of this use of the drug is lineally derived from him is also readily demonstrable. His immediate inspirer is evidently — from his quotations — Dr. Sydney Ringer; and whence did he learn it ? In January, 1869, he wrote a paper on Aconite in the *Lancet*, which begins — " Of all the drugs we possess, there are certainly none more valuable than aconite. Its virtues by most persons are only beginning to be appreciated, but it is not difficult to foresee that in a short time it will be most extensively employed in the diseases immediately to be noticed." And these are, inflammation and its accompanying fever. Dr. Ringer implies that he is making a new departure in the employment of the medicine, and the silence of the Materia Medicas of the day substantiates the assumption. He is equally silent as to the source of his fresh knowledge ; but the homœopathic history of the drug fills up the gap he has left. I venture to think that my own attempt to make our Materia Medica intelligible and interesting to outsiders had some share in this result ; for my *Manual of Pharmacodynamics* first appeared in the summer of 1867, and of course contained the statements which, eighteen months later, Dr. Ringer so emphatically warranted.

That it was the homœopathic method — small dose and all — which first revealed the antipyretic virtues of Aconite, and that it is homœopathic literature from which all knowledge of it has been derived, is thus the plain witness of history.   But what of Dr. Bartholow's further contention, that "Aconite is an antagonist to the fever-process," that "it is not applicable in accordance with the so-called law of similars"?   To answer this question we must ask a previous one, — what is the "so-called law of simi-lars"?   It is — as Hahnemann always laid it down — *similia similibus curentur*, let likes be treated by likes.   It assumes nothing about processes or antag-onisms : it simply requires correspondence — as close as possible — between the phenomena of disease and of drug-action.   *How* a remedy, prescribed on such indications, acts within the organism has always been left an open question ; and the majority of those who have thought about it, from Hahnemann himself downwards, have regarded it as probable that, appar-ently homœopathic, it is really antipathic to the essential morbid process.   Let it be granted that Aconite so acts in fever: how does it make against the claim to it of the method of Hahnemann, unless it can be shewn that the drug does not cause fever — is not febrigenic?   Dr. Bartholow might even deny this, as he is of course ignorant of homœopathic literature, and does not seem acquainted even with the experiments of Schroff.   The latter would give him some pretty evident febrile symptoms ; but the Austrian provings of our own school, from milder and more repeated doses having been taken, exhibit the result far better.   Here is one of them, which I extract from a monograph on the drug by Dr. Dudgeon in a forthcoming volume to be issued by the Hahnemann Publishing Society : —

"A. B., a healthy peasant girl, very robust, aged 22, took on successive days 5, 10, and 20 drops of tincture without effect." This was of course the Hahnemannian tincture of the whole plant, which is comparatively a weak one. "On the 18th January took 30 drops ; confusion of head, feeling of heat in the evening. 19th. — 40 drops ; after two hours confusion of head, sometimes changing into heavy feeling and pressive pain in crown and forehead ; loathing, nausea, general malaise with painful heaviness in limbs. After five hours pressive pain in scrobiculus cordis, dry feeling in mouth, great thirst. Felt so unwell, head so confused, giddy, and painful, and limbs so heavy, that she had to go to bed. Face hot, hands and feet cold, pulse contracted, hard, moderately quick. After nine hours, pressive pain in forehead, face turgid red, eyes sparkling, dry feeling in mouth, tongue moderately moist, slightly furred, no appetite, great thirst, oppression of chest, shallow quick breathing, with frequent deep breathing and sighing. No pains, but heaviness and fulness in chest, anxiety and palpitation of heart. Heart's beat strong, pulse fuller than usual, hard and strong, at same time moderately quick, skin warm, urine clear and reddish. After twelve hours heat and restlessness increased, tosses about from one side to another. In right thorax some pressive pain. After sixteen hours (9 P.M.), more tranquil the last hour, general warm sweat: headache, throat, and breathing difficulties relieved; pulse large, soft and slow. Only confusion of head and perceptible beating of heart complained of." Next day she was nearly herself again.

This — which finds several parallels in the Austrian provings — should be sufficient; or, if any distrust observations on the human subject, they may be referred to Dr. Mackenzie's experiments on animals recorded in the *Practitioner* of 1878-9. He found Aconite always to increase the temperature until asphyxia set in — the thermometer in the ear of a rabbit rising from two to four degrees Fahrenheit under its influence. Aconite is undoubtedly febrigenic; and that its curative action is homœopathic is also shewn by the small doses with which it may be effected. Hahnemann was content, as we have seen, with the 24th dilution; and though most of us now-a-days prefer to go lower in the scale, yet who can suppose that the thousandth part of a drop of the juice can act as an "arterial sedative," while in such dosage (the 3rd decimal) the drug is most promptly febrifuge in cases suitable for its administration.

And now, what are these cases? When Hahnemann spoke of the "pure inflammatory fevers" as its sphere of action, it is obvious that he did not mean so much the fever symptomatic of local inflammation, as that which from its resemblance thereto was named "inflammatory," under whatever circumstances occurring. For he specifies, as instances of its efficacy, measles, purple-rash, and pleurisy — the third only having any local affection as its basis. The Aconite-fever is thus essentially a synocha, and whenever this type presents itself the drug is indicated. It may occur in the exanthemata, after passing a catheter, or as the result of cold; but always and everywhere Aconite is its great remedy. It is rarely if ever present in the fevers we call toxæmic, as resulting from some morbid poison; and in the treatment of these — in typhus, typhoid, variola, and such like, Aconite plays but a small part. The same may be said of such blood-poisonings as septicæmia and pyæmia. As regards the fever symptomatic of local mischief, — when this is of a character to induce hectic, Aconite certainly finds no place. But what of pure inflammations and their accompanying pyrexia? Hahnemann commended it for these where "with thirst and rapid pulse, an anxious impatience, an unappeasable restlessness and an agonised tossing about are conjoined." Carroll Dunham has well shewn that these are the symptoms present when an inflammation is yet quite incipient. When substantive local changes have occurred, the tension of the circulation and nervous system diminishes, and any subsequent fever is sympathetic only. This Aconite will not touch : its place, as Teste says, is "in phlegmasiæ primarily general, and only secondarily localised;" and in these — I may add — only before the localisation is complete.

These canons about the relation of Aconite to inflammatory fever may be well illustrated by the

example of pneumonia. The pyrexia of this disease
seems quite independent of the phlegmasia: it precedes any manifestation of the latter by physical
signs, and subsides while these are yet in full presence. Aconite would therefore appear to be thoroughly suited to its incipient stage; and there are
some, both in our own school and in the other, who
maintain that — given in time — it can abort the
disease. But careful clinicians like Jousset and
Kafka warn us against indulging such expectations,
and my own experience is entirely in accord with
theirs. The acute pulmonary attack which Aconite
so promptly resolves is *congestion*, — a condition
which never goes on to croupous exudation, but —
if not arrested — leads to œdema pulmonum and
death. It is very rapid and dangerous, and to have
such a remedy for it as our present drug is a great
cause for thankfulness. But it is not pneumonia.
The rigor with which this malady generally sets in
indicates that the lung is already inflamed, and has
heated the blood up to shivering-point. You may
not yet hear any thing abnormal on auscultation and
percussion; but if you will count the respirations,
you will find that they are rapid out of all proportion to the pulse. When this is so, let me urge you
to waste no time in administering Aconite, but to
give Bryonia or Phosphorus, Antimonium tartaricum or Iodine, according to the patient's symptoms
and condition. You will under these find pulse and
temperature subside, gradually indeed, but with all
reasonable rapidity, and that from the very first;

while the physical signs diminish almost *pari passu.*
If twenty-four hours are thrown away in Aconite-
giving, you are more likely to have the sudden de-
fervescence seen in cases left to expectancy (which,
as the late Professor Henderson shewed, is due to
compression of the pulmonary circulation by the
filling up of the air-cells) with slow resolution of
the exudation.

Putting all these things together, and remembering what we have learned to-day concerning pyrexia
in general, it would appear that the sphere of Aconite is the nervous system of the circulation ; and
just so far as a fever belongs to this portion of the
organism, so far can this drug induce it in the healthy
and remove it in the sick. It is antipyretic, not by
diminishing the hyperoxidation on which (ordinarily)
depends excessive heat-production, but by regulating the apparatus provided for heat-liberation. It is
in the fevers brought on by a chill, and in whose
hot and cold stages alike the skin is dry and the cutaneous vessels tense, that it displays its greatest
powers. And hence, on the one hand the rapidity,
on the other the short duration, of its action. When
once the tension of the nervous and circulatory sys-
tems has been relaxed, and the pent up heat liberated,
Aconite has nothing more to do ; but this admits of
being done within a very short time, and Aconite can
do it. We do not continue this remedy for days
together, as with Belladonna and Arsenic. Hahnemann's single dose, or the frequent repetitions of the
later practice of most of us, accomplish their work in

twenty-four hours at the utmost ; and then, if necessary, other remedies come in.

The only other antipyretic which can be classed
with Aconite is *Gelsemium*, for Veratrum viride —
which some suppose analogous to it — seems to me
to belong rather to the Belladonna group. Of the
reputation with which Gelsemium came to us from
the "Eclectic" school of this country its power over
fevers formed no small part. Dr. Douglass, its earliest
prover, obtained results which shewed its action in
this sphere to be of the homœopathic kind. Experimenting on some seventy persons, he found it not
unfrequently to induce, in a few minutes, a marked
depression of pulse, with chilliness especially along
the back, cold extremities, and heat of head and face,
with pressive headache. This was soon followed by
a glow of heat and prickling of the skin, with full
pulse, rising as much above its normal standard as
before depressed below it ; and then came perspiration, sometimes profuse and lasting from twelve to
twenty-four hours. In reply to an objection that
others had not been able to produce these symptoms, Dr. Douglass wrote that the degree of chill and
febrile re-action in his subjects bore a very uniform
ratio to the nervous sensitiveness of the patient. In
some, where this quality was very marked, "the chill
was equal to a respectable fit of the ague, the re-
action and pain of head corresponding, and the sweat
profuse." [1]

Thus also must be explained the negative results

[1] Hale. *New Remedies.* 2nd ed., p. 409.

of Drs. Ringer and Murrell, in their recent provings, with regard to fever ; though they acknowledge, in one-third of their observations, a quickening of the pulse ranging from six to twelve beats in the minute, and in one experiment a rise of .2° to .4° Fahrenheit in temperature. The drug is certainly not antipathic to the febrile state, and yet its virtues therein are largely acknowledged. Dr. Douglass's statements favour the idea that it is in neurotic fevers that it is likely to find its best employment — in those which start with heat-confinement, and subside with heat-liberation. There is a remittent fever often seen in childhood, but not absent from adults, arising from various causes, but marked by this, — that the heat is almost absent in the early part of the day, increases towards evening, and subsides *without perspiration* as the night wears away. Gelsemium is as effective in this pyrexia as Aconite in its own. Even in the latter, it may vie with its prototype when, instead of anxious restlessness and craving thirst, there is rather a torpid and heavy condition, with a pulse not very rapid and inclined to be full and soft. The Gelsemium fever is a synochus rather than a synocha, — the main symptoms being those of languor and oppression, with dark crimson face, and dull pains in head, back, and limbs, the head feeling large and full. On this it acts with much rapidity, speedily relieving heat, oppression and aching. You may meet such a pyrexia as the result of a chill, in influenza, and in the milder exanthemata. It may also characterise some of the less pronounced malarious

fevers, especially in imperfect convalescence after
the intermittent paroxysms have been broken by
quinine.   Whether Dr. Hale's recommendation of it
in hectic can be sustained, I am unable to say.

At our next meeting we will take up the groups
of antipyretics headed by Belladonna, Arsenic and
Bryonia respectively.

VI.

PYREXIA AND THE ANTIPYRETICS (*continued*).

II. In 1838 Dr. Graves, the celebrated clinician of Dublin, advocated the use of *Belladonna* in those cases of fever with cerebral disorder which are attended with contraction of the pupil. He supposed that the mydriasis caused by it was due to its action on the brain, and that a cerebral condition accompanied by myosis must be precisely opposite in character. Hence, on the principle *contraria contrariis* — always influential with though often repudiated by our old-school colleagues — he argued that it ought to be beneficial in such conditions ; and supported his contention by several successful cases. His paper on the subject may be read in the *Dublin Journal of Medical Science* for July in the year mentioned.

We now know that Graves' assumption was incorrect ; that Belladonna does not dilate the pupil through the ocular nerve-centres, but by a peripheral action. The condition of brain it sets up would be accompanied with contraction of the pupil, were it not for the local influence it exerts on the ciliary nerve-terminations in the iris. To idiopathic cerebral states, there-

fore, of which myosis is a feature, it bears the relation of similarity instead of opposition ; and Graves
was practising homœopathy without knowing it.   A
man like Pereira felt there was something wrong
about the proceeding, and justly argued that — on
accepted principles — Belladonna should be contra·
indicated in febrile and acute inflammatory cases.
This seems to have been the general feeling, and
Belladonna found no place in the treatment of fevers
for another thirty years.

In 1869, however, Dr. John Harley published his
Old Vegetable Neurotics, and one of the drugs studied
therein was Belladonna.   In regard of its relation to
fever he speaks of "the similarity of the general
phenomena which attend its operation and those
which accompany pneumonia, enteritis, the development of pus in any of the tissues or organs of the
body ; " and says — "an infinitesimal quantity of atropia — a mere atom — as soon as it enters the blood,
originates an action which is closely allied to, if it be
not identical with, that which induces the circulatory
and nervous phenomena accompanying meningitis,
enteric, or typhus fevers."   These statements are
based on experiments shewing that the drug causes
a decided increase in the force and frequency of the
circulation, dryness of the mouth and tongue, an elevation of temperature, and an increase of the urinary
solids.   One would have expected that, under these
circumstances, the presence of fever would have been
to Dr. Harley, even more certainly than it was to
Pereira, a contra-indication for Belladonna.   It is just

the other way.   In the therapeutical chapter, the first
four maladies mentioned as under the control of the
remedy are pneumonia, enteric fever, typhus, and
acute nephritis.   When we come to enquire how he
could be led into such practice, we find, for the local
inflammations, a theory that the Belladonna dispels
the stasis by its stimulation of the sympathetic fibres,
and consequent narrowing of the capillaries ; but as
regards the fever, whether secondary or primary, all
pretence at antipathy is given up.   " Two similar
effects," he writes " the one arising from a local irritation, and the other from the presence of belladonna,
like spreading circles on a smooth sheet of water,
interfere with and neutralise each other ; " and again
— " it appears that the stimulant action of belladonna
is converted in great measure in febrile diseases into
a tonic and sedative influence."

This was somewhat bare-faced homœopathy to be
professed by a hospital physician and Gulstonian lecturer ; but it is characteristic of the advance of medical thought that no exception was taken to it.   I am
not aware that the practice has gained much acceptance ; but it is mentioned by all writers on pharmacology and therapeutics, and no one objects to it on
the score of its homœopathicity.   The febrigenic
power of the remedy is admitted by all ; and Dr. de
Meuriot, in his *Étude de la Belladonne*, relates observations shewing its power to raise the temperature
in dogs from 1 to 4 degrees of the centigrade scale,
and in man from $\frac{1}{2}$ to $1\frac{1}{10}$.

In the school of Hahnemann, this property of the

drug has long been recognised and turned to account. In the first edition of the *Reine Arzneimittellehre* (1811) the master gave numerous extracts from cases of poisoning which shewed fever as resulting from it ; and the provings of his associates which swell the pathogenesis in the later editions exhibit many symptoms of the same kind.   Hahnemann himself, as we know, was led by the principle of similarity to employ it in one febrile state as early as 1799, viz. : in scarlatina.   He does not speak of any other in his subsequent writings ; but the obvious application must have been largely made, for in Hartmann's *Practical Observations on some of the chief Homœopathic Remedies*, published in Germany in 1839, and translated here by the late Dr. Okie in 1846, Belladonna is given a high place in the treatment of many forms of fever.

We have now to enquire what these forms are ; to what fevers it is suitable.   And here our attention is at once drawn to the marked action of Belladonna on the nervous centres.   It is one of the "narcotics" of old, of the "neurotics" of present nomenclature : the sensorium is readily disordered by it, and delirium is among the most prominent of its poisonous effects. The inference is that its fever is due to hyper-oxidation of nervous tissue, and this is substantiated by the fact that it is the *phosphates* of the urine which specially shew increase under its influence.   Here is a pathological thought to guide us ; and it will lead us, as the symptomatic phenomena of the drug led our predecessors, to fevers of the typhous kind as especially calling for its use.   The "typhus nervosus" of the old nomenclature is its special sphere, best when this is also "versatilis," but not excluded (as Hartmann has shewn) when it may be called "stupidus." Of the essential fevers now recog- nised, the typhus and enteric types are those in which it plays its chief part; and it finds here a symptom very characteristic of it — the dry tongue. This is one of the most readily-induced physiological effects of Belladonna; and Dr. Harley has shewn it to be coincident with rise of the pulse, and to be replaced by a peculiar moisture on its fall. It is itself, moreover, no mere deficiency of secretion from inertness of glands or occlusion of blood- vessels. The parts are dark-red and congested, and the mouth is hot : we have (again to quote Dr. Harley) "a condition which exactly resembles that accompanying the typhus state."

There is another feature of the Belladonna-pyrexia which suggests the forms of fever to which it is appli- cable, and this is the rash on the skin so often ob- served. It is usually compared to that of scarlatina, and, with the fever itself, the sore-throat, and the delirium, it makes the drug's action a singularly close analogue to the disease. You know how this similarity was turned to account by Hahnemann, in proposing Belladonna as a prophylactic against scar- latina; and the success of the practice is vouched for in quarters far removed from his influence. In the malady itself, in its ordinary forms, the medicine should always be given; though I find it hard to say what is the precise amount of control exerted by it.

Much more striking is its action in the initial fever of variola, where also, every now and then, erythema-tous and scarlatiniform exanthemata precede the distinctive pustular inflammation, as Simon and Curschmann have shewn.[1] Aconite has no influ-ence here; but Belladonna can produce a decided abatement of the pyrexia and pains, even before the pocks come out. This is vouched for by physicians of both schools; and I can add my mite of confirma-tion to their testimony.

I have no desire, in these lectures, to make any pronouncement on the subject of dosage. But I must point out that confidence in Belladonna as an antipyretic is most felt and expressed the less atten-uated the form in which it is given. If you read Hartmann's essay (who used it at about the 30th potency), you will find it rather coming in to meet special indications in the course of fevers than to control the entire pyrexia. Those who give from the third dilution downwards have learnt to trust in it as a medicine which gives them real power over the febrile process. It does not act rapidly, as Aconite does, — the nature of the fevers to which it is appli-cable does not allow of this; but steady persever-ance with it will give the most gratifying results.

With Belladonna are to be classed its two sister-drugs — Hyoscyamus and Stramonium, and, more remotely, Agaricus and Veratrum viride. Let me say a few words upon each.

1. *Hyoscyamus* appears to cause fever as Bella-

[1] Ziémssen's *Cyclopædia*, vol. II.

donna does, by setting up hyper-oxidation in the nervous centres. In doses insufficient to dry the mouth and tongue, it actually lowers the pulse; but when that full effect of it is obtained, "the pulse" (writes Dr. Harley, who has experimented on this "old vegetable neurotic" also) "will generally ex-perience an acceleration of 10 to 20 beats, and be increased slightly in force and volume." With this the face will be flushed and the head oppressed; while the urine, if examined, will shew an increase in the urea, the phosphates and the sulphates, as with Belladonna, but not so marked. A reference to the "Fever" section of Allen's *Encyclopædia* will manifest general heat of surface as no uncommon feature of poisoning by the drug.

From Hahnemann onwards, Hyoscyamus has been esteemed by us a valuable remedy in "cerebral typhus." To indicate it in preference to Belladonna, the head symptoms should be those of oppression rather than of excitement, though there may be hallucinations or (if the patient is unconscious) inco-herent mutterings. The curious condition known as "coma vigil," where the patient is wide awake and yet insensible to surrounding objects, has been pro-duced by it and will generally indicate it. In similar states occurring in other fevers, as the puerperal and (as I have just lately seen) that of measles, Hyoscy-amus will give every satisfaction.

2. Of the power of *Stramonium*, in poisonous quantities, to cause fever, we have evidence as good as for that of Hyoscyamus; and, its influence on

the nervous system and other parts being so very
like that of Belladonna, its *modus operandi* as febrigenic may fairly be presumed identical. As compared with the latter, it causes more functional
excitement of the brain, but less active determination
of blood thereto. Its delirium, indeed, approaches
to mania; and on this very account it is less frequently indicated in fevers than the other mydriatics.
Where, however, its more intense nervous erethism
is present; where there are convulsive movements,
trembling, restlessness, loquacity, emotional agitation, you may substitute it for its congeners with all
confidence.

3. That *Agaricus* is febrigenic, we are hardly warranted in affirming. Some apparently febrile chilliness was observed by Lembke, in his proving on
himself; and some heat symptoms are reported by
two of Hahnemann's provers. On the other hand,
the very thorough experimentation of the Austrian
Society failed to develope any thing of the kind.
Roth rested his commendation of it in ataxic typhus
on other grounds. "The uncommon weakness" he
writes "and bruised feeling of the muscles of the
back and lumbar regions, the meteorism, the rum-
bling, the diarrhœa, the prostration of strength, the
stupefaction, the delirium, the small pulse, the trembling motions of the extremities, and many other
symptoms of the central nervous system, exhibit
great similarity to the ataxic forms of typhus fevers."
The indication has been abundantly verified, both in
this country and in my own: I, too, can bear witness

to its soundness. Tremor, loquacity, restlessness
and constant desire to get out of bed, are the indications for it ; while the delirium and unconscious-
ness of Stramonium are absent.

4. Of *Veratrum viride* a somewhat similar account
has to be given. It is not for us to use it as an
arterial sedative : the advocates of *contraria contrariis*
may do their best (or their worst) with it after this
manner. Experimentation on animals shews that
over and above the actions which conduce to this
result it has a direct influence on the nervous centres, — one of its alkaloids, viridia or jervia, causing
convulsions demonstrated to be of cerebral origin,
and post-mortem investigation showing intense capil-
lary congestion within the cranium, especially about
the cerebellum and pons. In the human subject
choreiform muscular contortions and paroxysms of
opisthotonos have been observed from over-dosing,
and much pain was felt by the provers in the frontal
region and the neck. It is where, therefore, in fever,
erethistic and hyperæmic conditions of brain and
cord occur that the drug should prove homœopathically useful ; and so it has been. It appears to be
of real value in the incipience of the epidemic cerebro-spinal meningitis which sometimes invades your
country, and also to be specially suited to the condition of the nervous system and the circulation in the
lying-in woman. Dr. Ludlam, one of the best of
our authorities here, speaks of its " wonderful power
to control and to regulate the vascular movements, to
equalise the circulation, and, as it were, to stamp out

a local congestion that would almost inevitably result
in inflammation," when occurring in the puerperium.
"It restores the milk and lochia, when these have
been suddenly suppressed, quiets the nervous per-
turbation, relieves the tympanites and the tenesmus,
whether vesical or rectal, and frequently cuts short
the attack." The tongue-indication Dr. Hale gives
for it — red-streaked down the centre, with yellow
sides — has been warranted by practice; but the
hard pulse of which he speaks is precisely opposite
to that physiologically produced by the drug.

III. In my last edition of my *Pharmacodynamics*
I have said the best I can say of *Arsenic* as an antipyretic. I have traced its history in this capacity,
shewing that only in the malarious fevers has its
remedial power been discovered by the empiricism of
traditional medicine, and that even here its present-
day revival has come from homœopathic sources;
while any application of it to the hectic and typhous
forms of fever is homœopathic *ab initio*. I have
shewn that it is unmistakably febrigenic, — inducing in some persons that peculiar morbid condition
of the nervous centres which shews itself in recurring febrile paroxysms (which are often periodic); in
others, a persistent febrile state which has been several times compared to and even mistaken for typhus.
In virtue of the former action, I have claimed for the
law of similars its undoubted power over ague, and
have urged therefrom its usefulness in such hectic
phenomena as characterise tuberculosis and pyæmia.
On the strength of the latter, I have propounded it

as the great remedy, not only for typhus itself, but whenever the well-known "typhoid" symptoms occur — the prostration, the pallor, the fuliginous tongue and lips, the (often involuntary) diarrhœa. Whenever these phenomena appear, I have said, whether in the continued fevers or the exanthemata, whether as symptomatic of mortification or as results of blood-poisoning, Arsenic is our main reliance, and should be used freely and persistently.

In these doctrines I have only given expression to the *consensus* of the homœopathic school. Connecting the facts with our present line of thought, I think we may say that Arsenic excites fever by acting on both factors of the process, — by disordering the nervous control over the temperature, and by increasing heat-production in some portion of the body. Looking farther in search of this portion, I think I shall carry you with me in suggesting it to be the blood. The power of Arsenic to destroy the life of the circulating fluid has long been inferred from the phenomena of poisoning by it, and now has been established by computation of the number of the corpuscles before and after its influence is set up. Hyperoxidation of no ordinary degree is here apparent, and there must be a corresponding rise in temperature. We take our stand on these facts in claiming for homœopathy the recent employment of the drug in pernicious anæmia (which, I may note, is generally accompanied with febrile phenomena): we may also use them in interpreting its action as an

antipyretic.   It is above everything *toxæmia* which
indicates it in fever : in proportion as the

> " life of all the blood
>   Is touched corruptibly "

is its control exerted.

The use of Arsenic in typhoid conditions is a piece
of homœopathic practice which has not yet been —
what shall I say? " the wise *convey* it call " — not
yet " discovered " by the followers of traditional medicine.   Dr. Ringer, however, has made a beginning
of its employment in the hectic type of fever in
which he will doubtless find many followers.   (He
had been to some extent anticipated in France, as
by Trousseau and Isnard ; but I speak of England.)
" It is stated " he writes in the tenth edition of his
Therapeutics " that arsenic will reduce the temperature in tuberculosis, and after carefully investigating
this subject, I am inclined to believe so ; at least I
have frequently observed a steady and sustained fall
of the thermometer follow the use of arsenic in cases
where the undue temperature had continued unchanged for a considerable time, and this I have
known happen twice or three times in the same
case on reverting to arsenic after it had been discon-
tinued.   The decline generally takes place gradually,
and may begin soon after taking the drug, or the fall
may be postponed for ten or twelve days."   The
gradual diminution of the pyrexia, here noted, is an
important and valuable feature in the action of our
present drug, and applies equally to the typhoid

types of fever. It produces no sudden deferves-
cence, as the cold bath and the large dose of quinine
or digitalis may do. These act on the high temperature solely, leaving its cause untouched; and
their effects are as evanescent as they are rapid.
Arsenic goes down to the very furnace where the
preternatural heat is being generated, and there
gradually but permanently extinguishes the fire.

The double action of Arsenic, — corresponding
on the one hand to such fevers as the malarial, the
hectic of phthisis and marasmus, and the pyæmic, on
the other to those of typhoid character, — this, I
say, gives it two kinds of analogues. In the former
sphere, it finds allies in bark and its alkaloids; in the
latter, in the serpent-poisons.

1. The relation of *Cinchona* bark to the malarious
fevers is still a moot one; and the question has entered on another stage during the last ten years. On
the one side evidence has accumulated in favour of
malaria consisting in a *contagium vivum :* on the other
quinine has been proved to be one of the most potent
protoplasmic poisons we possess. The inference is
natural, that it cures ague by destroying its cause;
that it checks the multiplication in the blood of the
germs whose inhalation has caused the mischief. If
this be so, the rationale of its action is the same as
that suggested by Hahnemann for the power of camphor over cholera; and we have an explanation of the
substantial dosage which either remedy seems to
require. Veratrum and Cuprum will play their part
in cholera even in high attenuation, while camphor

must be given in its primary solution and gains (as Dr. Rubini has shewn) by making this a saturated one. Those who treat intermittents with potencies seem never to use quinine and rarely its matrix bark, which Hahnemann maintained to be homœopathic to and declared to be specific for any recent malarious fever occurring in otherwise healthy persons. The inference is that their dosage is at fault; and that by using the drug in more appreciable quantities they would get as good results with it as others claim.

But the analogy extends yet farther. Camphor-poisoning exhibits cold and collapsed conditions very like those of cholera: it is therefore truly homœopathic to the condition there present, and may benefit otherwise than as a germicide, as well as in similar states occurring independently of the Asiatic scourge. And so it is with bark. In the work to which I have referred I have shewn by numerous citations from unquestioned sources, that cinchona and quinine have the power of exciting in the healthy the febrile paroxysm. Whatever, therefore, they may do in intermittents by killing the malarial bacilli, so far as they influence the living tissues it is in a direction similar to that of the disease, and thus (as we know) antidotal to it. They may accordingly cure agues — as they often must have done — without acting as germicides at all; and they will be applicable to similar paroxysms not owning malaria for their source. In this way they take place as anti-hectics. This is the type of fever liable to be set up when suppuration is proceeding, or any great drain

on the system is going on, and this — in homœopathic practice — is the recognised sphere of our
"China." It is only a step farther when we come to
the pyæmic and septicæmic fevers. Quinine may
possibly do something here as a represser of the too
exuberant cell-life which is overwhelming the organism; but I am sure that it also acts, in doses too
small for such purposes, as a supporter to the invaded
system. In such cases its combination with Arsenic,
in the form of Chininum arsenicosum, is a valuable
remedy, — the hæmatic action of the Arsenic rein-
forcing the purely neurotic one of the Cinchona.

2. Here, however, we have come to toxæmic fevers once more, and to the other branch of arsenical
analogues, the serpent poisons. The use of these
in medicine is peculiar to homœopathy, and constitutes one of its most potent instruments. It should
be so according to the law of similarity, for what is
more quickly fatal, more widely destructive, than
the bite of a venomous snake? and practice has confirmed the inference. In *Lachesis* and *Crotalus* —
the venoms respectively of the lance-headed viper
and the rattlesnake — we have antidotes to the worst
fevers that afflict humanity; and the names of those
who have done most to emphasise their value or supply materials for their employment, the names of
Hering, Neidhard, Dunham and Hayward, will (I
venture to predict) stand on lasting record.

The phenomena of snake-bites would suggest that
it is in malignant local inflammations, with secondary
blood-infection and nervous prostration, that serpent-

venom would play its part as a remedy; and this indeed has proved the great sphere of Lachesis. Traumatic gangrene, malignant pustule, dissection-wound, and all cases of pyæmia and septicæmia owning a local source, come within its range and respond to its action. But experiments in inoculation have shewn that the rattlesnake venom, at any rate, is capable of setting up a primary pyrexia, most analogous to the scourge of your Southern States, the yellow-fever. The brilliant comparative success which homœopathy has always scored in the treatment of this disease — and never more so than in the last epidemic — has been largely due to Crotalus. But the analogy leads us to apply it to purpuric febrile states in general, whether occurring primarily and ordinarily, as in epidemic cerebro-spinal meningitis, or constituting a distinct variety or occasional form of other maladies, as scarlatina and variola. Dr. Hayward's monograph on Crotalus, which will shortly appear, is likely to give a great impetus to this use of the medicine. He can say — "for the last ten years the present writer has used Crotalus in all cases of fever of all kinds when any thing of a hæmorrhagic or putrescent character has been exhibited, and with the most prompt and marked effect in removing the hæmorrhagic symptoms and checking the tendency to putrescence." His two cases of malignant scarlatina (to which even the *Lancet* has given insertion) well illustrate its power to rescue from almost imminent dissolution.

IV. We are now well-furnished with medicines which cause fever by wasting the nervous centres and the blood ; but we want another set which shall serve us when the morbid process is seated in the more lowly-organised tissues, in those of the vegetative system. The type of such medicines is *Bryonia*.

The earliest pathogenesis of this drug appeared in the second volume of the first edition of the *Materia Medica Pura*, i.e. in 1816, and it contains — especially in the portion furnished by Hahnemann himself — some marked fever-symptoms. It is probable that he had observed these some time previously, for as early as 1813 he had been led to give Bryonia as one of the chief remedies in the epidemic fever which ravaged the army that had fought at Leipsic. On reading his account of the matter (which you will find in Dr. Dudgeon's translation of his lesser writings), you will see that the symptoms which — in addition to the common febrile ones — led him to its choice were "shooting (or jerking-tearing) pains in the head, throat, chest, abdomen &c., which are felt particularly on moving the part." These express its action on fibrous, serous and muscular tissues. When delirium supervened — when, that is, the hyperoxidation involved the nervous centres — Hyoscyamus had to take its place.

Now it was obvious that if pains increased on movement were to be the characteristic of the Bryonia fever, it should be above all things a remedy in acute rheumatism ; and so it has proved to be, as we shall see in our next lecture. But if, so indicated, it could control the whole febrile condition present, it must evidently play some part in the treatment of the essential fevers, of which that treated by Hahnemann was an instance. It was possibly (as Dr. Russell thinks) of the kind now known as "relapsing," though I do not find any mention of relapses as occurring in it.[1] The success he reported with it, and some additional symptoms given in the second and third editions of his *Materia Medica Pura*, led to its taking among his followers a high place among the remedies for the common endemic "typhus," which was generally that which we now call typhoid or enteric fever (their "typhus abdominalis"), but sometimes the "synoque" of the French, the "gastric" or "common continued" fever of our English nomenclature. All the representative writers of the first period of homœopathy — Hartmann, Wolf, Trinks — concur in stating that the fever of Bryonia is a synochus, often so active as to simulate a synocha, and so (wrongly) to call for Aconite, but liable, if not checked, to go on to a typhus nervosus. Such a condition — vascular and nervous erethism, without much prostration — we often have at the commencement of true typhoid, but still more in the initial stage of the common continued fevers. To these it is probable that (as with Baptisia) Trinks' statement applies that typhus may sometimes be arrested with Bryonia. The features in them which have led to the appellation "gastric" are further indicative of our present medicine ; for such symptoms as coated tongue, foul and bitter taste, nausea and vomiting,

---

[1] In this "famine fever" Dr. Kidd, who saw so much of it in Ireland in 1847, considers Bryonia the best medicine.

and so forth, — which have not come before us in the fevers of Aconite, Belladonna, and Arsenic, are always strong indications for Bryonia. Let head- ache and other pains be further marked, and a ten- dency to involvement of the chest appear, and we have a pyrexia which Bryonia will arrest if arrestable, and at any rate will modify so favourably as to earn the patient's grateful thanks.

The allies of Bryonia are Rhus, Phosphoric and Muriatic acids, and Baptisia.

1. From their first joint appearance in the volume of the *Materia Medica Pura* issued in 1816, Bryonia and *Rhus* have ever gone together ; and these "great twin-brethren" have come to our aid in many a doughty fight. Rhus also had been found to excite decided febrile symptoms, not merely as sympathetic with the dermatitis caused by its external application, but primarily, from ingestion in moderate doses. Hahnemann accordingly administered it in the fever I have already mentioned, either after Bryonia, or from the outset, when the pains were worse during rest, and when greater prostration was present. Whatever this fever was, of 183 cases treated by him in Leipsic not one died; and so Rhus, like Bryonia, became a fever-remedy in the homœopathic school. Hahnemann himself resorted to them, as the standard febrifuges, in the consecutive fever of cholera, apparently recommending their alternation therein. In the common continued fevers — gastric and typhoid — the question has generally been which of these should be given ; and Rhus has been recog-

nised as suitable to an erethistic condition more
adynamic than that of Bryonia, but less so than that
of Arsenic. The first supervention of diarrhœa
upon constipation, of a red upon a coated tongue,
calls for it in gastric fevers ; and in true typhoid may
often indicate it from the first. "Comparing it"
writes Dr. Dunham " with Bryonia and Eupatorium,
we miss at once the whole train of gastro-hepatic
symptoms, the vomiting of bile, soreness and pain
at the pit of the stomach, constriction around the
epigastric zone, fulness and tenderness of the hepatic
region, etc., which indicated those remedies in bilious remittent fevers. On the other hand, we find
Rhus producing some degree of tenderness of the
abdomen, great flatulent distension of the abdomen,
amounting to tympanites, occasional watery or mucous diarrhœa, — symptoms which, though not so
strongly pronounced as similar symptoms are under
Phosphoric acid, yet decidedly resemble the symptoms of typhoid fever, and indicate the use of Rhus
in that disease." A red triangle at the tip of the
tongue has recently been said to indicate it here.

2. *Phosphoric acid* was reckoned by Wurmb and
Caspar, the distinguished clinicians of the Leopold-
stadt Hospital in Vienna, an analogue of Rhus in
typhoid, corresponding to symptoms of the same
degree of severity, but marked by torpor rather than
erethism. How they came so to use it I cannot tell ;
for there is little indication of fever in its pathogenesis, and neither Hahnemann nor any of his earlier
disciples mention it in this connection. Dr. Dun-

ham, however, as we have seen, seems to think the indications warranted ; and they are confirmed by the high value set upon it by the no less eminent Paris clinician, Dr. Jousset. "Phosphoric acid" he writes "is, with Muriatic acid, one of the principal medicines for typhoid fever in its stage of full development : abundant diarrhœa with involuntary stools, paleness of the stools, but most of all intestinal hæmorrhage, indicate the remedy. As concomitant symptoms we observe epistaxis, bleeding of the gums, pallor of the face or one cheek red and the other pale."

3. With Phosphoric goes *Muriatic acid ;* and, to my thinking, outdoes it far in range and energy in the febrile sphere. Our employment of it here must indeed be traced to tradition ; for the fever-symptoms of Hahnemann's pathogeneses of the drug are quite insignificant. The tradition took its rise from iatro-chemistry : the chlorine and the acid were supposed to neutralise respectively the putrescence and the alkalescence of the fluids in low febrile states. That the action was really dynamic appears from the high esteem in which it has been held in the homœopathic school, when given in doses far too small to exert any other kind of influence. The tendency to ulceration of mucous membrane observed in cases suitable for it suggests that it is where the hyper-oxidation of fever affects chiefly this tissue that it is most applicable ; and it gives it a special relation to typhoid. Putrescent phenomena, with us as of old, are always indicative of it ; and it thus fre-

quently finds place in the treatment of malignant scarlatina and diphtheria.

4. Last, but not least, I have to speak of *Baptisia*. The question of the power of this drug to abort true typhoid is, I think, definitively closed. There are occasional instances of this fever which spontaneously terminate at an early period, and any one who uses Baptisia much for it will every now and then meet with such a case. If he ignores the numerous examples of the malady in which no such result has followed, he may become an advocate of its abortive power here. But whenever a succession of cases of true typhoid has been submitted to the treatment, the result has always been failure.

There is, however, a fever in which Baptisia is specific, and whose course it does positively arrest. This is the "gastric," or common continued fever of which I have already spoken, — a fever which runs no definite course, but which, under ordinary treatment, mostly puts on a typhous appearance before coming to its close. That Baptisia is the Aconite of this fever, that it can break it up in perspiration in (at most) two or three days, is a thing of which I am as certain as of anything in therapeutics. It was the confusion of this fever with true typhoid which led to the baseless claims some of us (among whom I must include myself) have made for the remedy; but we must not, in the re-action, abandon it altogether. Such a malady is often seen, is sometimes epidemic, — it constitutes the "colonial fever" of Australia and the Cape; and the fame of Baptisia in the past, and its solid usefulness in the future, lie within the sphere of it.

With Baptisia, as with Muriatic acid, putrescent phenomena are characteristic; and fœtor of the breath and secretions alike indicates it and yields to its use. It may thus play a part in other fevers where such a feature is apparent, in scarlatina, in diphtheria, and above all in variola. Dr. Eubulus Williams' experience with it in the last-named, during an epidemic occurring in a children's home at Bristol, was strikingly favourable; and he especially notes the absence of the usual offensive effluvia in the cases treated by it.

We have yet to learn what it is in suppuration which causes hectic, and it may well be some process allied to putrescence. Dr. Mitchell, of Chicago, has praised Baptisia highly in this fever when occurring in connexion with phthisis; and I have verified the indication when the tongue has been moist and coated, instead of — as in the cases calling for Arsenic — red and apt to become dry.

These, ladies and gentlemen, are the leading classes of our antipyretics, and the drugs that fall within them. The groups now formed and comparisons instituted will aid you, I trust, to apprehend more clearly their genius and sphere of usefulness, and so to employ them with enhanced precision in the various febrile states which come before you. Rightly used, they will give you — I believe — such control over fever that you will need none of the harsher

measures of the old school to supplement them. A
temporary exception must be made for the sudden
hyperpyrexia of which I have spoken, where according to our present knowledge the cold bath or pack
is indispensable for the saving of the patient's threatened life. Save under these rare circumstances,
however, homœopathic medication, with ordinary
coolness and quiet, abundant drink, and simple regimen, will form the best of antipyretic treatment.

VII.

RHEUMATISM AND THE ANTI-RHEUMATICS.

IN taking Rheumatism as the second peg on which
to hang a bundle of medicines, I am aware that I am
assuming that to be an entity which pathology refuses
so to recognise. The numerous affections known as
" rheumatic ", ranging from acute polyarthritis to a
simple stiff-neck, have — it is said — neither a common specific cause nor a single anatomical seat : already a number of them have been differentiated as
gout, rheumatoid arthritis, myalgia, &c., and the remainder hold their place provisionally only. But,
whatever may be said theoretically to such criticism,
it still holds good practically that around the true
" rheumatic fever " which all agree is an " essential "
malady there are grouped a number of affections more
or less closely allied to it, owning (mostly) the same
atmospheric causation, occupying similar parts, and
— which to us is of most importance — responding
to similar medicines. I maintain that these are as
much entitled to the term " rheumatic " as is the
febrile polyarthritis specially so named, and I shall
include them as varieties of the malady of whose
remedies I have now to speak.

By rheumatism, then, we mean first of all the acute general form — " rheumatic fever." As the name implies, fever is a marked feature here : it even seems to precede the affection of the joints, though — as in pneumonia — the appearance may be deceptive. It is admitted by all to be proportionate to the number of the joints involved and the intensity of their inflammation. It has the curious feature of being accompanied almost from the outset by free perspiration, which, however, affords no relief ; and the sour odour of the sweat is known to all who have attended a case of the kind. It is in acute rheumatism that the temperature is most apt to run up to that menacing height in which it is known as " hyperpyrexia." This, however, must not be regarded as an excessive degree of the true rheumatic fever, which is an inflammatory one : it is confessedly rather of neurotic origin, and referable to a collapse of the heat-controlling centres in the cerebro-spinal system. The inflammation is primarily situated in the synovial membrane of the joints, — though it may spread to the neighbouring parts ; and it is characterised by great instability, suddenly supervening and as suddenly departing, the (purely serous) effusion being as rapidly poured out and re-absorbed. Certain secondary inflammations are very apt to occur during the course of the disease, viz. : peri- and endo- carditis and pleurisy, less frequently meningitis and pneumonia. That the serous membranes should be specially affected their close alliance with the synovial would make appropriate ; but it is to be noted that when inflammation attacks these in the course of rheumatic fever, it is not fugitive as in the joints, but takes a firm hold, and — especially in the heart — often leaves lasting results.

The cause of this disorder is in most cases a chill, less frequently a fright, occurring to persons heated by exertion; in either case checking perspiration. It is reasonably inferred that its phenomena are due to the retention in the blood of excrementitious products, capable of setting up local inflammations and general fever; and that the sweating of the disease is an eliminative effort, its odour being due to the acid nature of the *materies morbi*. In the rare cases in which rheumatic fever developes spontaneously, we may suppose that the same products have . resulted from mal-assimilation or imperfect retrograde metamorphosis.

Whether this theory be true or not is of some therapeutic importance. In the first place, it rules the application we shall make of Dr. Balthazar Foster's two observations (confirmed by one of Kuelz's) that the administration of lactic acid — at any rate to diabetic patients — is liable to set up acute rheumatic arthritis in various places; while Richardson and Rauch have found it, in animals, capable of inflaming the endocardium. By our law, lactic acid should be a prime remedy for rheumatic fever, if it were not — as it is by the hypothesis just stated — the very *materies morbi*. Certainly no reports of success with it have appeared in homœopathic literature; and their absence favours the other alternative.

If it be true, the present supplies illustration of the impossibility of always working with symptomatic resemblances only, and ignoring pathology.

But then, if lactic acid in quantity be the pyrogenous and phlogogenous matter of rheumatic fever, is not chemical antidotal treatment — as by alkalies — indicated therein? It is difficult to resist the inference theoretically, and practically it has been maintained that not only is the whole duration of the malady lessened by such treatment, but — which is more important — the frequency of cardiac complication is materially abated. The objection to such treatment lies in its flooding the system with the necessarily large quantities of alkali required,[1] — an evil somewhat mitigated by substituting salts of soda for those of potash hitherto given, but generally resulting in greater anæmia than is natural during convalescence. The total outcome of experience with alkalies has not hindered our brethren from grasping eagerly at the prospects held out by salicin and the salicylates, nor has it stood in the way of trimethylamine or dethroned colchicum. These last are remedies which act like our homœopathic specifics, striking, as we may suppose, at the cause rather than neutralising an effect. For, although the starting-point of rheumatic fever may be the retention of lactic acid in the blood, from checked perspiration, its persistence for weeks must imply (as such a product has no self-multiplying power) a continual

---

[1] " No effect can be expected from alkalies unless they are given in very large doses " (*Senator*, loc. cit.).

fresh formation of the *materies morbi;* and here we must hypothecate a morbid protoplasmic activity, by the cessation of which alone can the disease be radically cured.

In adults, at least, we are therefore warranted in relying solely on specific measures. In young people, viewing the much greater frequency of heart-affections, and the apparent efficacy of alkalies to avert them, their administration would be desirable, were we not capable of achieving as good results with purely homœopathic treatment. It would also appear, moreover, that it is only in comparison with other perturbative treatment — as by blisters, opium, &c. — that alkalies secure this superior immunity; since cases left to nature under good nursing do as well. I am not sure, however, whether the experiment has been carried out on a sufficiently large scale, and whether — as with Dietl's in respect of pneumonia — a greater number of cases might not alter the averages.

But we must return to our clinical survey. Chronic articular rheumatism is allied to rheumatic fever by being not uncommonly its *sequela;* and, when it is of primary occurrence, by its resulting from the same causes, viz. : continued exposure to cold and damp. It is more fixed in seat, and leads to greater changes in the substance of the joints : otherwise, it may be described as consisting of the local affection of acute rheumatism without the constitutional disturbance.

" Muscular rheumatism " is a good deal discredited now-a-days ; and, when not of systemic origin (as in scurvy and metallic poisoning), is set down as either myalgia or myositis. I cannot think this procedure wise. The true "myalgia" ought — I think — to be restricted to muscular pain coming on after too prolonged, sudden, or great exertion ; while myositis can hardly be present without tenderness of the inflamed muscle, — a symptom of which in the acutest cases of lumbago there is hardly any trace. The majority of instances of this malady, of pleurodynia, and of torticollis, can be traced to the main causes of rheumatism generally, and may fairly therefore be considered "rheumatic" in their nature.

Again, the fibrous tissues throughout the body, especially the fasciæ and nerve-sheaths, are, under similar circumstances, liable to acute or chronic morbid conditions whose pain, stiffness, and sensitiveness to atmospheric changes assimilate them to rheumatic arthritis. They too may therefore be considered part of our disease-group, and will be found to have a very defined relation to certain anti-rheumatic remedies.

To these we now proceed.

I. *Bryonia*, of all medicines, comes nearest to being a typical analogue of rheumatism, in almost all its forms. It is a specific irritant ; and the parts it inflames when absorbed into the blood are, not so much skin and mucous membrane, as serous and synovial membrane, fibrous tissue, and muscle. It has excited, especially in one of the heroic Austrian provers (W. Huber), decided fever ; and in several of them the urine was of febrile character. Hahne-

mann and Otto Piper have each noticed sour sweat
from it. As regards the joints, a number of the
symptoms of its pathogenesis (see S. 1345, 1425,
1426, 1550 and 1551 of Allen's *Encyclopædia*) speak
of swelling, stiffness and tenderness in them ; and
a very instructive experiment was made on himself
by Dr. Elias Price, of Baltimore, who, after taking
doses of the tincture ranging from 30 to 50 drops,
had — after five days' incubation — a decided attack
of rheumatism of the foot, not entirely passing away
until seven weeks had elapsed. For the muscles,
we have soreness and pain on motion experienced
by the provers in many parts of the body, even to
the production of pleurodynia and lumbago ; and in
an autopsy made on an animal poisoned by it, the
substance of the heart and the muscles of the neck
were found intensely red.

When the rheumatic significance of the pathogenesis of Bryonia was first noticed, I do not know.
Hahnemann does not intimate it ; but Hartmann,
in 1839, speaks of it as already "a very celebrated
remedy." Curiously enough, popular practice had so
far anticipated its applicability to joint affections as
to give it the name of *Gicht-rübe*, gout-root. Whatever it may do for the malady referred to in this
appellation, there is no doubt that in the treatment
of rheumatic affections of all kinds it plays a most
important part. In rheumatic fever, some of us
alternate it with Aconite throughout — Dr. Yeldham
advising the one to be given during the day, the
other at night : some prefer to give Aconite alone

until it has done all it can do, and then to put in our
Bryony.   There are two indications, however, which
should lead us to administer Bryonia by itself and
at the first.   One is pathological: it is when the
disease presents the "synovial" as distinct from the
"fibrous" form.   This distinction was made by
Chambers and Francis Hawkins, and has been en-
dorsed by Watson.   In the synovial variety there is
more local swelling and less general fever: as the
name implies, the synovial membranes of the joints
are the seat of the inflammation rather than the
tendons, fasciæ, and ligaments round about them;
and we have seen that it is these chiefly on which
our medicine acts.   The other indication is symptomatic : it is where the sensitiveness to movement
so characteristic of the drug is markedly present.
In proportion to its degree not only is Bryonia called
for, but its dosage becomes more attenuated, so that
from the 12th to the 18th, or even higher, it displays
the most brilliant powers, as testified to — *inter alia*
— by Bayes and Madden.   I mention this, because
generally in rheumatic affections the lowest dilutions
seem to do most good.   In the complications of
rheumatic fever, Bryonia is quite competent to deal
with the pleurisy and the pneumonia ; but when the
heart becomes affected it yields place to more definitely cardiac remedies.   In chronic articular rheumatism Bryonia occupies a sphere which may be
exhibited in the words of Sir Thomas Watson : —
"There are two kinds of chronic rheumatism : one
attended with local heat and swelling, although the

constitution at large sympathises very little or not at
all with the topical inflammation ; the other charac-
terized rather by coldness and stiffness of the joints.
In the former of these the pains are increased by
pressure, and by movements of the limbs, and by
external warmth ; the warmth of a bed, for example ;
and there may be even some slight degree of pyrexia
at night.   In truth this form of chronic rheumatism
claims a near relationship with the acute, into which
it sometimes passes, and of which it is frequently
the sequel." I may anticipate so far as to say that
the great remedy for the second variety here described is Rhus, but for the first Bryonia is specific.
In acute muscular rheumatisms it also plays an important part, though in severe lumbago I confess I
prefer Aconite. The pain on movement of our medicine is of course present here to a marked degree ;
and in pleurodynia where it is indicated it obliges the
patient to lie on the affected side. For the chronic
fibrous rheumatisms Bryonia can do little. It is to
maladies caused by dry cold, such as that brought
by the east winds of my country, sweeping across
the arid steppes of Russia, and traversing no ocean
of any breadth which can temper their harshness,
that this remedy is appropriate ; while the affections
in question trace their origin rather to cold associated with damp.   .

Bryonia, in its relation to rheumatism, may fairly
be regarded as the head of a group of medicines ;
the other members of which are Aconite, Colchicum,
Pulsatilla, and the product variously known as "pro-

pylamine" and "trimethylamine." To these we will
now direct our attention.

1. Next to Bryonia, our great anti-rheumatic is
Aconite. Here we have not been, as with the other,
the first to establish its virtues. Stoerck, in 1750,
experimenting on himself with the extract, found it
to cause considerable diaphoresis. He inferred that
it should prove useful in chronic local rheumatisms,
and found it so to be, the improvement generally
setting in with sweating and eruptions on the affected
parts. From this the transition to acute local rheumatisms and at length to rheumatic fever was natural: Tessier (of Lyons), Lombard and Fleming were
its agents, and you will find in Stillé several other
testimonies to its value. Nevertheless, the remedy
has not occupied the place which might have been
expected in ordinary practice ; and I agree with Dr.
Phillips that this has resulted mainly from the strong
doses in which it has been given. The aim has been
rapidly to hush up the pain in the joints or to knock
down the fever : this the drug can only do as an
anæsthesiant and arterial sedative, and its effects
when thus given are so unpleasant and even dangerous that its use comes to be abandoned.

That so it is appears from the steady favour ac-
corded to it in the school of Hahnemann, where
gentle and unperturbing dosage is practised. We
believe it to be homœopathic to acute rheumatism,
and use it accordingly — absorbing its whole physiological in its therapeutical action. We conceive that
we have good grounds for so regarding it. In relation

to fever I have brought before you only the neurotic
action of Aconite; but it must not be forgotten that
it is also an irritant. You will sometimes hear it
said that Aconite does not cause inflammation; and
this is true if the process in parenchymatous organs
is intended. But in *post-mortem* examinations of
animals poisoned by it decided evidences of inflammation of the pleura and peritoneum have been
found; and the sensations of some of the provers
point in the same direction. Similar pains, of cutting, shooting and tearing character, were experienced by them in the fibrous tissues generally, the
muscles, and the joints, — those in the last-named
having been observed also in cases of poisoning.
One of Schroff's provers of the Aconitum camma-
rum experienced them to a high degree. In no case,
however, is there — as with Bryonia — any redness
or swelling.

Having regard to these facts, and to the relation
of Aconite to the synocha present in this disease,
we should say that it ought to be the prime remedy
in the "fibrous rheumatism" of Chambers and
Hawkins, where the peri-articular tissues are mainly
affected and the fever runs highest. This indeed it
is. You must not use it as a neurotic in this instance, and expect it to subdue the fever in twenty-
four hours. It cannot here help us by liberating the
arterioles from their contracted state, and promoting
heat-radiation. This is going on abundantly, but
heat-production is on its side so greatly enhanced
that no relief is gained thereby. In rheumatic fever

Aconite is rather antiphlogistic than antipyretic : it acts on the protoplasm which is being too rapidly oxidised, and for this time is required.

It is the testimony of all who have trusted to Aconite in acute rheumatism that under its use the heart is much less liable to be affected. Should it be so, however, you need not intermit the use of the drug ; for its homœopathicity and activity here have been abundantly proved. "One of the Austrian provers had, alternating with his articular sufferings, painful palpitation of the heart and præcordial anxiety; and Dr. Jousset says that he has introduced into the circulation of rabbits increasing doses of the extract, with the invariable result of producing lesions of the mitral valve" (*Pharmacodynamics*). If Aconite is actually being taken when the cardiac complication supervenes, it may be alternated with another heart-medicine, as Spigelia ; but otherwise it may be given alone, and relied on alone.

In our dosage, Aconite has hardly found place in the treatment of chronic rheumatism, articular or fibrous ; and in acute muscular rheumatism it generally gives way to Bryonia or Cimicifuga. I would make an exception, however, in the instance of lumbago. It caused a real attack of this nature in Schneller, a physician of the old school who proved it at Vienna in 1848, and I have always found it very effective as a remedy. Here, however, as in rheumatic affections generally, you must not — I believe — attenuate the drug too far. I seldom rise above the 1st decimal dilution, and generally give drop-doses of the tincture.

2. The second of the analogues of Bryonia is
Colchicum. The repute of this medicine has mainly
been acquired in the treatment of gout; but Watson
is not the only physician who declares of acute rheumatism, that "the preparations of colchicum have
sometimes an almost magical effect in subduing the
disease." When we enquire how they act, we find
that it is by no indirect process, for it is quite unnecessary to induce the purgative operation of the
drug: nor is it by any antiphlogistic power, since in
ordinary inflammations — not being gouty or rheumatic — it is quite inert. The action is direct and
specific; and to account for it we can only look to
the physiological effects of the plant. Let us take
a case of poisoning by it, reported in the *London
Medical Gazette* for 1838–9. A woman, at 33, took
nearly an ounce of the tincture of the seeds one
Thursday at 11 P.M. Symptoms of gastro-enteric
irritation soon supervened; but among these, about
10 A.M. on the Friday, she felt a numbness of the
feet and hands. "To this succeeded" writes the
reporter "a pricking feeling, as if, so she expressed
it, they had been asleep. All the joints of the
fingers and toes, and also the wrists and ankles,
were very painful, and the toes and fingers were
painfully flexed at times. Pain in the shoulder-joints
succeeded, and, on Saturday, in the hips and loins.
It was also increased in intensity, so that she said
she thought she should go mad. Ultimately, almost
all the bones and joints were affected with pains,
which were of a gnawing, dragging character. Just

before these symptoms were at the height, very profuse sweats came on, and were of a very sour odour.
I may mention, that to the pains in the limbs were
added, on Sunday morning, great stiffness and pain
in the back of the neck and occiput, which was
aggravated by moving the head : there was also a
feeling as of something tightly bound round the
head ; and on moving the tongue, and in swallowing,
she experienced sharp pain about the root of the
tongue. *She asked me if she had not got rheumatic
fever*. . . . The pain in the joints continued exceedingly severe long after the vomiting and purging had
ceased : it was still so on Tuesday."

These effects are paralleled at points in several
of the numerous provings which have been made
with the drug, both in the old school and in our
own ; though here perhaps the pains have been felt
more in the muscles than in the articulations. Two
other observers, however (Krahmer and Lindemann),
have noticed the sour sweats so characteristic of
rheumatism ; and in a boy who had eaten portions
of the plant the left elbow and knee-joints were
swollen, hot and painful. . The numbness and pricking of the hands and feet observed by Mr. Henderson
(the reporter of the case of poisoning) is interesting,
as it is a well-known effect of Aconite, and we shall
see it again in Propylamine.

After all this, I am sure that any candid enquirer
would bear me out in claiming for homœopathy anything that Colchicum can do in rheumatism. Mr.
Henderson very justly observes that the symptoms

of his patient would suggest that it was not by its purgative effects — as then supposed — that the drug proved remedial, but by its specific action on the parts affected. We have not used it much, as in Bryonia we have a remedy so closely resembling it, and taking its place in that very "synovial" rheumatism to which it has hitherto been allotted. But the poisoning I have cited would suggest it as hardly less applicable to the "fibrous" form, if another medicine seems needed after Aconite has spent its force. As with the two preceding anti-rheumatics, it has shewn a specific influence on the serous membranes by inflaming the pleuræ; and it has displayed remarkable power in controlling pericarditis when occurring in the course of rheumatic fever. It should be more used than it is in muscular rheumatism, especially torticollis, as it seems to have an elective affinity for the muscles of the neck. It caused pleurodynia in several of its provers; and in a gouty patient of Dr. Russell's proved strikingly curative of it. Its pathogenesy bears Teste out in saying that the rheumatic pains to which it corresponds are generally *tearing*. He adds that in warm weather they are principally felt at the surface of the body : as the air grows cooler they seem to penetrate the deeper tissues and the bones.

3. The third drug to which I would draw your attention in this category is *Pulsatilla*. It is one which is still peculiar to homœopathic practice, though our esteem of it in orchitis has lately received corroboration from several practitioners of the opposite camp.

Its action on the joints was manifest in Hahnemann's provings of the nigricans species, and appears also in those which have been made in this country with the Pulsatilla Nuttalliana. In most of the articulations the pains, however severe, were transient; but in the knee, and especially in the feet, they were accompanied by swelling, indicating synovial effusion.

The general action of Pulsatilla would indicate its use in sub-acute rheumatism, of synovial type, with little or no fever; where the knees, ankles, or small joints of the feet and hands are affected; and when the articular inflammation shifts readily from place to place. The temperament and disposition of the patient will in such rheumatisms generally be found that characteristic of Pulsatilla, and — when present — will lend additional force to its indications. Its relation to the digestive organs, moreover, would point to it as specially applicable when faulty assimilation rather than chill had evoked the disease. In chronic articular rheumatism the indications are the same; and here its "conditions" must be specially regarded, viz.: that its pains are worse towards evening and at night, worse also at rest and in a warm room, and relieved by motion in the open air. These features of the action of the drug we owe to Hahnemann; and they are invaluable in determining its choice. On the fibrous tissues and muscles Pulsatilla has no action.

Another relation of this medicine which must be remembered is that which it bears to the generative organs in women. There is an ally of rheumatism and gout, specifically distinct (however) from either, variously known as "rheumatic gout," "nodular rheumatism," or (better) "rheumatoid arthritis." When fully developed, no arthritic disease is more hopeless and more disabling : the best chance for its victims is to take it in the sub-acute form in which it not uncommonly begins. Now Dr. Fuller has noted that "in early life, rheumatic gout is always hereditary *or connected with disordered uterine function;*" and I would add that, in later life, its most frequent subjects are women at the climacteric period. Pulsatilla finds an obvious sphere of action here ; and has more than once justified its choice.

4. The last drug of our present group is *Propylamine.* I use this name for the present, as it is that by which the substance in question is known in commerce, under which it was first introduced into medicine, and by which it is designated in homœopathic literature. It is more properly, however, trimethylamine. Methyl, not propyl, is its radical, — three parts of this displacing three parts of the hydrogen of ammonia, the resulting body being expressed as C_3H_9N. The substance used in medicine is the propylamine of commerce purified, — this in its turn being obtained from many animal products, of which herring-brine is one of the best known and least unpleasant. It is not pure methylamine — some ammoniacal salts almost inevitably cleaving to it ; but the former can be made synthetically, and — given as a chloride — seems to act much in the same way as that obtained analytically. As these prod-

ucts are little known, I may mention that Propylamine dissolves freely in ether, alcohol, and water;
and that, if you have to give it strong enough to
make it desirable that its disagreeable taste should
be disguised, you may accomplish this to some degree
by adding a little peppermint. The chloride of trimethylamine is equally soluble, but inodorous and
not repulsive to the palate.

Propylamine was first introduced into medicine
about 1855, by Awenarius, a Russian physician, who
— after using it in 250 cases of acute rheumatism —
reported it as little less than specific. Giving two
drops only every two hours, he "frequently saw all the
symptoms of the disease vanish after twelve doses
had been taken." Dujardin-Beaumetz, in France,[1]
and W. H. Spencer, in England,[2] have since reported
results almost equally good; and the remedy was
just coming into favour when the salicylic treatment
— of which I shall speak immediately — dethroned
and extinguished it, so that it finds no place in the
last edition of Ringer's or of Wood's *Therapeutics*.

It is not our way so to lose a potent medicine, and
it has never been forgotten in the homœopathic
school. We have studied it in our usual manner,
i.e. first pathogenetically, by observation and experiment, and then therapeutically. Observations on
patients treated by it have yielded little definite
result as regards its physiological action. Dujardin-
Beaumetz, however, proved the chloride of trimeth-

---

[1] *Bull. Gén. de Thér.*, LXXXIV., 227, 337, 395.
[2] *Practitioner*, Feb. and Mar., 1875.

ylamine on himself and another healthy person, and in both the force and frequency of the pulse were reduced, and the temperature fell nearly or quite one degree centigrade. One of our own colleagues, Dr. Chaffee, of Kentland, reported to the Indiana Institute of Homœopathy, in 1880, a proving on himself. " I took " he says "ten drops of propylamine in water. One half hour after taking the same I experienced a smarting sensation of the tongue and fauces, with much thirst : there was also a tingling of the fingers, a sensation of numbness to such an extent that in attempting to pick up any thing it felt heavy, and I had to use great effort to retain the article within my grasp. This group of symptoms passed off at the expiration of three hours. I then took another dose of ten drops and experienced the above symptoms intensified, with the addition of great pain in the wrist joints, also great restlessness, yet inability to stand upon my feet from the pain produced in the ankle-joints. I took no more of the medicine, the tongue became broad and flabby, the mucous membranes of the buccal cavity were pale, appetite gone, no desire for anything, became morose, with great desire to be let alone : the pain in the joints was made worse by the slightest movement. Twenty-four hours after taking the drug, I was attacked with diarrhœa : the stools were thin, watery, and white." [1]

Several casual experiences with the drug have been reported in our journals, and one of these may

[1] *U. States Med. Investigator*, June 15, 1880.

be cited as worthy to stand beside those from old-school sources. Dr. Sanborn, of Hampton Falls, N.H., writes — "In January, 1862, I began to use Propylamine in rheumatic fever, and have used it every year since. My whole number of cases has been about sixty. I begin the treatment with Aconite and Propylamine, giving two drop doses of the latter four times a day. I have had no case which lasted more than a week, and generally in three days my patients can walk about without pain."[1] This very fairly corresponds with the experience of Dujardin - Beaumetz and of Spencer. The latter of these, I should say, notes that the remedy does not produce such marked and speedy effects in lymphatic subjects as in those of nervous and sanguine temperament. It is therefore allied to Aconite rather than to Pulsatilla.

I can thus commend Propylamine to you as having all the features of a homœopathic remedy, acting directly and in small doses. I cannot say the same of the drug which I have mentioned as having superseded it in ordinary practice. I mean salicin, with its derivatives salicylic acid and the salicylate of soda. I call these last derivatives of the first, as — though ordinarily obtained from other sources — salicylic acid appears to be a product of the breaking up of salicin in the system, and the action of the two, physiological and therapeutical, seems to be identical. Now I do not deny that these substances may occasionally be homœopathic to acute rheumatism.

[1] *Medical Call*, July, 1882.

Both have been known to induce fever, which, though of short duration, has been pretty sharp; and the salicylate of soda has every now and then, in the hands of practitioners of our school, done exceedingly well in doses which the ordinary practice would deem inert. But I cannot claim for homœopathy the salicylic treatment of rheumatic fever as a whole. To be of any use — all who employ it affirm — the dosage must be large, — ten to thirty grains every two hours or so. When thus administered in any febrile disease, a considerable reduction of temperature is obtained, and in acute rheumatism the pain and inflammation of the joints is also greatly abated. But that the essential malady is not touched appears from the fact that cardiac and other complications are at least *as* liable to occur, and that relapses are decidedly more frequent than when improvement has resulted from other measures. As compared with alkaline treatment, for instance, heart-mischief is nearly twice as frequent, and relapses occur three times as often.[1] Still, it might be said, giving all due weight to these disadvantages, the benefit obtained is so great, and so unattainable by other means, that the salicylic treatment should not be withheld from our patients. But there is yet another demerit in it. The large dosage required is an evil which has not been sufficiently considered, either here or in the analogous instances of the bromide treatment of epilepsy and the iodide of syphilis. You cannot introduce these masses of for-

[1] *Lancet*, Sept. 20, 1879.

eign matter into the system without serious injury. Already men are beginning to trace kidney disease to the too free use of iodide of potassium, as the surgeons are finding out that carbolic acid poisons patients as well as germs, and the aurists are protesting against the injury done to the ears by the wholesale quinine-giving now prevalent. That salicylic acid and its salts are liable to similar reproaches is pretty well known. As early as 1877 it had to be said that "in a considerable proportion of cases they give rise to disagreeable symptoms, such as vertigo, headache, tinnitus aurium, and deafness, nausea and vomiting after every dose, profuse sweating, great weakness, and occasionally a peculiar eruption on the skin. More rarely, the symptoms assume a dangerous complexion, violent delirium, albuminuria, great prostration, with pallid skin and feeble pulse, ushering in fatal collapse." [1] Since then, necrosis (in a strumous child) of the bones of the legs and forearm,[2] hyper-pyrexia, and hæmaturia, are among the disastrous effects observed from this acid. Salicin is said to be exempt from these reproaches, but if, as Senator maintains, it is transformed into salicylic acid in the system, the mischievous agent is still produced, and — though less manifestly — does its injurious work.

I think, therefore, that on the whole we shall be doing most justice to our acute rheumatic patients if we resist the temptation to hush up their pain and

[1] Appendix to vol. xvi. of Ziemssen's *Cyclopædia.*
[2] *Lancet*, Oct. 27, 1877.

knock down their fever with salicin and its derivatives. A man must make his choice : he cannot
have every advantage and escape every draw-back.
Under homœopathic treatment his disease will subside somewhat less rapidly, but no less surely ; and
he will run no risk of being poisoned during its
course, or unduly weakened when he arrives at convalescence. If our patients, knowing what as a general thing homœopathy can do in comparison with
ordinary treatment, on this ground seek our aid, we
may very well be content to treat them homœopathically. I am sure that in the long run they will benefit thereby ; and that we shall do better by studying
to give precision and effect to our own anti-rheumatics, than by hankering after the flesh-pots of
Egypt in the shape of the medication now under
review.

VIII.

RHEUMATISM AND THE ANTI-RHEUMATICS
(*continued*).

II. AT our last meeting, after a review of the phenomena and pathology of rheumatism, in its various
forms, we discussed a group of medicines related to
it as remedies, of which Bryonia was the leading
member, and Aconite, Colchicum, Pulsatilla and
Propylamine the subsidiaries. We shall begin to-
day with a second group, the type of which is to be
found in a plant very familiar to you in this country
— the *Rhus toxicodendron*.

Some ten years ago, there appeared in England
a treatise entitled "Materia Medica and Therapeutics. Vegetable Kingdom. By Charles D. F. Phillips, M.D., F.R.C.S.E." The author was described
as "Lecturer on Materia Medica at the Medical
School of the Westminster Hospital." On turning
over the pages of this volume (which has since been
reprinted over here) any one could see that it contained a large amount of matter hitherto unfamiliar
in the old school ; and any one acquainted with homœopathic literature could recognise the source

from which the new matter had been derived. The fact was that the writer had been an avowed practitioner of homœopathy in Manchester for twenty years ; but on coming to London found it suit his views ostensibly to renounce his hitherto connexions. It was nevertheless the special knowledge he had acquired as a disciple of Hahnemann which enabled him to get into metropolitan practice, to obtain his appointment, and to write and publish his book.

I have made these remarks, because I am going to quote Dr. Phillips' description of those pathogenetic effects of Rhus which concern us in our present study. "Rhus induces pains," he writes, "apparently of a rheumatic kind, and which are felt not only in the limbs but in the body, though most especially about the joints. Pain and stiffness in the lumbar region are often induced, and to these affections is often added a sense of numbness in the lower extremities. The structures most powerfully affected appear to be the fibrous ones. The pains in question are accompanied by a very slight amount of swelling ; and, singular to say, they become intensified by rest and warmth."

Now where did Dr. Phillips get these symptoms? His whole description of the effects of Rhus purports to be taken from cases of poisoning by contact with or emanations from it, — several instances of which, he says, he has himself witnessed. But it is a curious fact that in none of the poisonings collated by Allen (and he has spared no pains to include all on record), whether from Rhus toxicodendron or from the venenata species, is there any mention of rheumatoid pains being felt. The whole stress of the plant's influence here seems expended on the skin and the cellular tissue. In the provings of Hahnemann and his associates, however, where the drug was chiefly taken internally and in moderate doses of the tincture, tearing pains, and especially sense of stiffness and aching, in neck, back and extremities, were frequently observed, with the absence of swelling and the conditions of aggravation noted by Dr. Phillips. The source of his information is obvious : and what shall be said of his candour ? what shall be thought of the reviewers who welcome such a volume and cast scorn on the authority it implicitly recognises ?

We, happily, are under no such morally-benumbing influences, and can freely and openly use good knowledge wherever we find it. That Rhus can cause rheumatoid pain and stiffness we first learnt from Hahnemann's pathogenesis of Rhus toxicodendron ; and we have had it confirmed by Dr. Joslin's proving of Rhus radicans, and those of Kunze, Burt and others made with Rhus venenata. In these last the joints, as well as the fibrous tissues, were affected — especially the knees, ankles, feet and hands ; but there was here also no synovial effusion, as there was with Bryonia and Pulsatilla. In all cases paralytic and numb sensations, with ready trembling, are present.

The repute of Rhus as an anti-rheumatic is of purely homœopathic origin, though its remedial

power is now beginning to be discovered in the old
school, and even turned to account in our doses.
At what point this repute originated among us, I
cannot say; for Hahnemann makes no mention of
rheumatism in his preface or notes to the pathogenesis in the *Materia Medica Pura.* We have seen,
however, that his main indication for the drug in the
fever of 1813, in which he so largely and successfully used it, was the presence of pains which were
worse at rest. This condition of aggravation of the
Rhus sufferings he mentions again with much empha-
sis in the preface referred to; and the inference
was natural that in rheumatic pains so characterised
it would prove as useful as Bryonia had been in its
way. Obviously, therefore, it would find little place
in rheumatic fever, where the patient generally
dreads the slightest movement. Cases will occur,
however, in which — instead of this feature — restlessness and constant desire to change the position
are present. Here Rhus will tell; and also in Aconite cases, after this drug has done all it can do, but
"wearing stiffness and aching in the neighbourhood
of the joints" (Phillips) annoy the patient. In
chronic articular rheumatism it is *the* remedy for
Watson's second form, where there is "coldness and
stiffness of the joints," and friction and gentle
motion relieve. In association with *dry* heat topi-
cally — as from hot bran, salt, or sand — it will
loosen many a stiffened and useless articulation. It
is especially valuable when this affection is chronic
from the outset, when the joints and fibres of persons

advanced in life begin to stiffen and ache, threaten-
ing complete and painful disablement.   My own
father, when 73 years old, began to be so afflicted ;
but under the steady use of Rhus, in rare doses of
the higher dilutions, he lost all his troubles of this
kind, and lived to 85 without any return of them.
But its main sphere is found in the local fibrous
rheumatisms of which I have spoken — this name
being due to them from their almost constant origi-
nation in getting wet.   It is this, as well as their
seat, which calls for Rhus as a remedy ; for another
point of distinction between it and Bryonia is that
the former (with — as we shall see — Dulcamara)
suits the effects of damp,[1] the latter (with Aconite)
those of dry cold.   It was for this reason that, in
speaking of Bryonia as remedial in the effects of
east winds, I explained that I meant those of my
own country.   Those which reach you here in Boston must bring plenty of Atlantic moisture with
them, and their effects will probably call for Rhus.

Well, then : for rheumatic pain and stiffness traceable to damp and increased by it, in tendons, fasciæ,
ligaments, Rhus is specific.   You cannot better
impress this action of it upon your minds than by
reading a series of cases by one of the early homœopathists, Dr. Bolle, of Dresden, which you will find

---

[1] Dr. Hoyne (*Clinical Therapeutics*, I., 128) quotes a case of pleurodynia
supposed to depend on cardiac disease.   It refused to respond to remedies
until traced to a wetting, when Rhus rapidly disposed of it.   This may be
compared with a similar case related by Teste (*Mat. Med.*, sub voce Aconitum), where exposure to dry cold winds on the Russian plains was found to
have been the exciting cause, and Aconite proved quite as effective.

translated from their German original in the twenty-fifth volume of the *British Journal of Homœopathy.* His conclusions are that Rhus answers to affections, occurring chiefly in men of strong fibre, occasioned by taking cold from wetting when the body was in a state of perspiration and excitement, but also when no such mental or physical strain had preceded; characterised by tension, lameness and stiffness, by tearing, drawing, bruise- and sprain- like pain in shoulders, wrist-joint, back, and vastus, and in the hips, and not unfrequently from those down the thighs to the feet, with occasional sensation of numbness. The pains had the usual aggravations and ameliorations, among the latter being dry heat.

With regard to the localities specified by Dr. Bolle, I would mention that you will often read or hear of lumbago and sciatica being relieved by Rhus, although it has no action on muscle or nerve. I believe it is the lumbar fascia in the first instance, the fibrous sheath of the sciatic in the other, which is the seat of the malady; and here of course Rhus is quite at home. It is thus well indicated when lumbago and sciatica occur together — the whole fibrous tract being affected. I would also remind you, before leaving this drug, that the indication "relieved by motion" is not so absolute as that of "worse at rest." I mean that the patient's first movement, after repose, may be painful enough : but he improves as he goes on, while the Bryonia subject gets worse the longer the motion continues.

The analogues of Rhus in relation to rheumatism are Dulcamara and Rhododendron.

1. *Dulcamara* has, since Carrére recommended it, found no little employment as an anti-rheumatic; though it has now pretty well died out of use, and Dr. Harley's experiments with it — which seemed to shew it totally inert — will not favour its revival. Its use among ourselves is not, I think, lineally traceable to that which it enjoyed in the old school. It has rather grown out of a phenomenon noticed by Carrére in some of the patients taking it, viz.: that they were liable to twitching of the eyelids and lips, and slight convulsive movements of the hands, *but only when exposed to cold, damp weather*, while external warmth readily removed the symptoms. It was a necessary inference that Dulcamara would be homœopathically indicated for other affections owning such causation; and if so then certainly for some forms of rheumatism. The inference was further supported by a considerable array of rheumatoid pains in Hahnemann's pathogenesis of the drug, and it has been warranted by experience. The following case, from the *Études Thérapeutiques* of the late Dr. Petroz, will illustrate its sphere of action:

"A young woman, of a delicate complexion, her skin white, fine, almost transparent, inhabited a room situated low down, where the temperature in summer differed from that of outdoors at least ten degrees" (centigrade).

"In 1833, during the hot season, every time that after moderate exercise she returned to her apartment, she felt an impression which, at first agreeable, finished by being unpleasant. One day she complained of heaviness in the head, with a feeling of stupidity; she had, at the end of the day, shaking in her limbs, some shooting pains in different parts, an extreme fatigue;

during the night wakefulness, or an agitated sleep, awaking with
fright, dry heat of skin, ebullitions of blood.

"The next day, besides the preceding symptoms, pulse hard,
contracted, 110; slight swelling of the articulations of the wrist
— carpus and metacarpus; sharp pain caused by the least movement; sensation of cold while resting: the articulations of the
feet were in the same state.

"In the morning, at 9, Dulcamara 24, three globules. At
noon some painful drawings were felt in the legs. At two a
general perspiration took place, without increase of heat, and
lasted all the night, which was more calm.

"The third day the skin was in its normal state; the pulse
was 88; touching and movement caused much less pain in the
affected joints. The improvement made progress.

"On the fourth day the patient was near convalescence,
when in the evening she felt violent shooting pains in the right
forearm and wrist, in the right thigh and knee; there was wakefulness and agitation.

"The morning of the fifth day: — frequency, hardness of
pulse; sharp pains in the articulations of the superior and inferior extremities on the right side. The analogy of these symptoms with the effects of Dulcamara was too evident not to lead
one to recur to that medicine: it was given in the same dose.
At the end of two hours the skin became moist: the fever was
soon calmed: the pains decreased progressively, in such a manner as to render movement possible.

"On the seventh day all pains had ceased, and since that
time the health has not been altered."

This is a fair case of incipient rheumatic fever
nipped in the bud by Dulcamara (which, I should
say, is not without febrigenic power); and it is possible that, had the dose been occasionally repeated
after the first, the recurrence of the fifth day might
not have taken place. You will see that quite a high

dilution was sufficient : it is so, in my experience, with this medicine, and also with Rhus, when either is indicated. You will see also that the Dulcamara pains have not the relation to rest and motion of those of Rhus ; and that the local affection is more synovial. They correspond in their sub-acute char-acter, and in their exciting cause.

2. *Rhododendron*, the Siberian rose, has a high native reputation for gout and rheumatism. Its proving by some of the earlier disciples of Hahne-mann excited sufficient pains in the muscular and fibrous tissues to shew that it was homœopathic to rheumatic affections of these parts, and it has been used accordingly with much success in homœopathic practice, — here also the attenuations given having been mostly those from the 12th to the 30th. Rheu-matism of the cervical and thoracic muscles, and rheumatic neuralgia of the extremities — arms and legs, figure most largely among the cases affixed to the pathogenesis, as originally published in Stapf's *Beitrage*. More lately it has acquired much repute for "rheumatic face-ache," i.e. pain about the cheek and jaws induced by a draught. The Rhododendron pains are like those of Rhus, in that they are worse at rest ; but they are relieved at once by movement. As the *habitat* of the plant suggests, they have no special relation to damp ; and they do not indeed re-quire cold for their exciting cause, though they might have it abundantly there. It is the electric condi-tion of the atmosphere to which they are especially sensitive ; and when you get the feature "worse

before a storm," you may always think of Rhododendron.

I am inclined to believe that Rhododendron actually affects the nerve-substance itself. At any rate,
its relation to rheumatism in the face and extremities
leads me to hang on to it, as a sub-class, two drugs
mainly efficient in the same direction, Kalmia and
Spigelia.

a. In his interesting preface to the proving of
Kalmia in the first volume of the Transactions of
the American Institute, the late Constantine Hering
connects the plant with Rhododendron and with
another anti-rheumatic which will shortly come before
us, the Ledum palustre. "Rhododendron" he writes
"thrives in the region of storms and mountains,
and Ledum draws its nourishment from the ponds
of elevated regions, while Kalmia flourishes in the
mists arising from the valleys. All these inhabit
northern climates. They correspond to the great
family of diseases which we comprise under the col-
lective names of rheumatism and gout; particularly
to that class which belongs to the north, and which
is decidedly distinct from that of the south, and of
the tropics." It caused rheumatoid pains in abundance in those who experimented with it. Dr. Bayes
used to esteem it highly in the rheumatic face-ache
of which I have just spoken, and Mr. Clifton confirms his estimate from his own experience, adding
that the pain is generally on the right side, and
often goes down the arm, while brachialgia itself of
similar origin will often yield to it. Kalmia affects

the heart much as Digitalis does, especially slowing its beat; and the affinity may be utilised when rheumatism attacks this organ. We should expect most from it when the heart simply becomes the seat of rheumatic pain, without inflammation of the organ being set up; but Dr. Dunham relates a case of endocarditis so completely cured by it that no valvular murmur remained.

b. *Spigelia* has a still more decisive action on the nerves and heart, and also on the eye. In the last-named organ the painful character of the inflammation induced in Hahnemann's provers suggests the sclerotica and iris as the parts affected, though the symptoms are not described with sufficient clearness to make a diagnosis possible. We may certainly say, however, that it is "rheumatic ophthalmia" which the drug causes, and may extend a similar conception to its influence on the heart, which is denoted by great pressure on the chest, shooting pains through it and down the left arm, and violent palpitations. Its neuralgia must not be limited to the rheumatic kind, as it is genuine and wide-reaching; but in practice an origin of this nature has always been found a strong indication for it. Its face-pains are sharper and more shooting than those of Rhododendron and Kalmia, and — unlike the latter — affect rather the left side.

In the eyes and heart, moreover, we must not limit Spigelia to rheumatic affections. Ciliary neuralgia of every kind, and angina pectoris when a pure neurosis, come within its range. But the *inflammations*

it controls in these organs are undoubtedly those of
rheumatic origin. The painful, straight-lined and
crimson-hued injection of the eye-balls which we
used to call "sclerotitis" yields rapidly to it, after
Aconite; and it stands *facilé princeps* among our
remedies for pericarditis and endocarditis rheumat-
ica. That Fleischmann should have relied upon it
year after year at the Gumpendorf Hospital in
Vienna for the cardiac complications of acute rheumatism, and should have been able to report 57 successive cases with only one death, is proof sufficient
of its efficacy.

III. I have now to speak of a group of medicines
whose seat and kind of action are "rheumatic," but
which are purely local in their influence. Rhododendron is the link which connects them with the
more generally-acting drugs we have left behind us;
but their own type is Actæa, or — as you seem here
to have determined to call it — Cimicifuga race-
mosa.

That *Cimicifuga* causes pains in various parts of
the body, and these often of some severity, no one
can doubt who reads its pathogenesis. That these
are — to some extent, at least — of rheumatic nature
appears from the interpretation which practice has
given them. The drug is praised by all in acute
muscular rheumatisms, as pleurodynia, lumbago and
torticollis — the first-named having been experienced
more than once by one of the provers. The differential features of its action here — as compared
with that of Aconite and Bryonia — are hardly yet

established. For its influence on local articular rheumatisms, however, we have such indications from an unusual source, viz.: from Dr. Ringer. It is useful, he says, in rheumatoid arthritis, and in joint-inflammations -resembling gonorrhœal rheumatism, but without any history of gonorrhœa, when the pains are worse at night and in wet or windy weather. The "uterine origin" which he requires for the former, if the drug is to act well, reminds us of Pulsatilla; and leads me to say that when pain and irritability of the womb, whether at the monthly period or otherwise, appear to be traceable to rheumatic influences, Cimicifuga is a great remedy for them. I know that present-day pathology would smile at such a conception; but it was entertained on sound data by men like Gooch and Dewees in the past, and I think we should lose therapeutically by abandoning it. That another hollow muscle, the heart, can be rheumatic is less disputed; and when it is so, its painfulness being less acute than with Spigelia, Cimicifuga gives it great help.

There are four remedies which hang on to Cimicifuga in its relation to local articular rheumatisms.

1. Its closest ally is *Caulophyllum*, whose uterine action is pretty well identical with it. Now this plant, when proved by Dr. Burt, developed in him pretty sharp pains in the small joints, and those of the fingers were stiff and red. It has thereupon been used with good effect in inflammatory rheumatism of the hands, and among its employers is Dr. Ludlam, who has noted its action as being more prompt in women than in men so affected.

2. *Ledum* is one of the old Hahnemannian medicines. Its proving developed marked pain in many
of the joints, and the ankles and feet appear as actually swollen under its use. It is perhaps better
suited to gout than to rheumatism ; but Pflange, one
of the Rademacherian school in Germany, reports
two good cases of chronic rheumatism of the hip-
joint cured by it. Hahnemann's *dictum*, that Ledum
will prove suitable only in chronic maladies characterised by coldness and deficiency of animal heat,
points to the non-inflammatory type of rheumatism
as its sphere, like Rhus, and like my next medicine,

3. *Ruta.* This, too, causes marked pain in the
joints, but goes beyond Ledum (which has shewn
some power of the kind) in involving also the neighbouring bones, probably acting on the periosteum. It
is thus, with the drugs of the Mercurius group to
which I shall immediately come, suited to the form
of rheumatism known as "periosteal ;" but is sometimes efficacious in the ordinary kind when affecting
the wrist, for which I think it has a special affinity.
You will not forget the uterine influence of this
medicine also.

4. Last of our present group in the violet —
Viola odorata. In the slight pathogenesis of this
plant which we have in the *Archiv*, furnished by
Hahnemann, Gross, and Stapf, two of the few symptoms of the extremities are "pressing pain in the
right wrist" and "drawing pain in the right elbow."
Petroz seems from this slight indication to have been
led to employ it, and Teste says with success, in vari-

ous rheumatic affections of the upper limbs.   Tessier then took it up, and published in the *Gazette Homœopathique de Paris* several cases of inflammatory rheumatism which rapidly recovered under this medicine in the 12th dilution.   In all the right side alone waś affected, and the wrist-joint was always the first to experience amelioration.   Dr. Kitchen, of Philadelphia, translating these cases for an American journal, adds three of his own, in which cure at least as rapid occurred under the 1st dilution.   In one the affection was limited to the right side, but in the other two it was bilateral; and here the curious feature appeared, that the left wrist seemed entirely insensible to the action of the remedy, so that it remained painful and swollen for some days after its fellow had got well, and compelled resort to other medicines.[1]

IV.   My next group of anti-rheumatics is a somewhat peculiar one.   It consists of three members only, and its type is *Mercurius*.

In the last edition of my *Pharmacodynamics* I have spoken (as previously) of the well-known "mercurial rheumatism," and have suggested that this is mainly a periostitis.   But I have also called attention to the profuse and odorous sweats giving no relief, which characterise rheumatic fever, and which — when occurring otherwise — are always indicative of the remedy; and also to the sallow and red tint of the face, with oily perspiration, which Dr. Anstie has remarked as frequently premonitory of it, and which

[1] See *Brit. Journ. of Hom.*, XXIV., 314.

no less plainly suggests mercurial preparations. I have accordingly recommended Mercurius to be given in sub-acute forms of the disease, readily relapsing, where the pains (unlike those of Pulsatilla) do not shift about much, and are markedly worse at night — the patient being very sensitive to cold.

I do not think, however, that we must lay too much stress on these symptomatic resemblances, so as to *depend* on Mercurius in rheumatic polyarthritis. Its fever (which no one has described more graphically than Hahnemann[1]) is very unlike the rheumatic, as contrasted with which also it has diminution instead of excess of fibrin in the blood. On the other hand, it shews undoubted power of locally inflaming the joints. From Huber's very exhaustive treatise on "Mercury and its preparations," now in the course of appearing as a supplement to the *North American Journal of Homœopathy*, I extract the following. "Dietrich thus describes the mercurial arthritis: A slight stitching, pressing pain in the joints sets in; the joint swells up and becomes of a pink or dark red color. This redness disappears on pressure, but returns immediately. The swelling is neither hard nor doughy, and feels hot to the touch. When at rest and at a low temperature the pain is moderate; but becomes more severe during motion and in the heat of the bed (especially from Merc. corr.). It may end in caries or anchylosis. Wilhelm, Handschuh, Warbeck de Chateau and Stokes also speak of a mercurial rheumatism, attacking the knee

[1] See *Pharmacodynamics*, p. 626.

and shoulder joint, more rarely the hip-joint, elbow and carpus; sometimes in the form of an acute rheumatism, leading to hydrarthrosis or to suppuration of the joint.   Richter ascribes especially to the corrosivus the power of causing rheumatitis; and Kussmaul acknowledges that painfulness of the joints is frequently observed in workers in mercury, but denies that inflammation or exudation in the joints follows such occupation.   Lendrick, on the contrary, observed, in consequence of the abuse of mercury, swelling of the joint with considerable exudation in its cavity."

Such facts go to warrant the recommendation of Mercury I have made in my *Therapeutics*, viz.: that it should be given when the inflammation is obstinate in any one joint.   They would also suggest it in a form of acute articular rheumatism which has not yet come before us — the gonorrhœal.   And this leads on to another "rheumatic" application of our drug.   You may be aware that there are some persons who never contract gonorrhœa without getting, not rheumatism only, but also *iritis*.   (This is quite different from the ordinary "gonorrhœal ophthalmia," which is due to some of the virulent matter coming in contact with the conjunctiva, to which membrane it is limited, and which attacks one eye only.)   Now the association of Mercury with iritis is quite traditional, and has descended without break of continuity into homœopathic practice.   In a paper which you will find in the tenth volume of the *Annals* of the British Homœopathic Society I have examined into the matter. I have shewn that the common idea that iritis is readily caused by Mercury is quite unfounded, — the two authorities usually cited in favour of it, Graves and Travers, both recognising in their cases the presence of two other factors of much greater importance, syphilis, and cold and damp ; while it has never been observed among the workers in the metal. A single exception made by Travers to his statement that all his patients were syphilitics, and an observation of Basedow's in which an "iritis mercurialis" appeared in a patient being treated with the drug for hepatitis, were the sole evidence I could find of the possibility of the disease being induced by it. Considering, then, the plastic nature of syphilitic iritis, and the necessity recognised by the most careful oculists for inducing to some extent the physiological effects of the drug if it is to overcome it, I was forced to the conclusion that it was not homœopathic thereto. Its reputation in our school was due, I thought, to its power over rheumatic iritis, especially when of serous character, and this variety of the affection it might have caused.

I wrote all this before Huber's essay had reached the section on iritis. I find here the usual assumptions made, which analysis readily resolves ; but a few fresh facts are contributed. "Many authors" we are told "such as Hunter, Bele, Scarpa, Pearson, Ammon, Werklin, Travers, M. Jæger and many others, do not admit the existence of syphilitic iritis, but ascribe the same to the effects of mercury." They may have been of this opinion, but if it rested on no better foundation than in Travers' case, their authority is of little value. If, moreover, supposed syphilitic iritis were really of mercurial origin, why did they all treat it with Mercury? The therapeutic evidence adduced by Huber goes against his thesis, as his two clinicians of most weight — Kafka and Payr — find it necessary, in the syphilitic form of the malady, to resort to mercurial inunction, as did Dr. Dudgeon in a case referred to in my paper. But three fresh observations, besides Basedow's, are given, which strengthen the evidence for the homœopathicity of the drug to some form at least of the disease. One is rather slight : — "Cooper, while administer· ing mercury in a scrofulous affection, observed the development of iritis." I do not know what Cooper this may be, nor where his observation is recorded ; so must let it pass *quantum valeat.* The second is taken from von Ammon's *Zeitschrift für Ophthal· mologie :* "In a man and a woman who had taken sub- limate for chancre, but were not salivated, there appeared inflammation of the right eye, which in- volved not only the posterior surface of the cornea, but also the serous covering of the iris. The pupil was angular ; the eye very painful, as if too small ; all the sensations aggravated in bed." Here the patients had chancre, which might have been syphi- litic ; but on the other hand the iritis was serous only, which makes against such an origin. Thirdly, the "case of Werklin" is mentioned, but no refer- ence given to it. Dr. Huber states that in this "iritis appeared together with gummy tumours in a patient who never had syphilis, and who took calomel with opium for facial rheumatism. In the list of enumerated symptoms" he goes on "we find injected state of the vessels; pressing, burning pain in the eye; photophobia; induration of the iris; irregular shape of the pupil, and hypopion; there was violent burning pain in the supra-orbital region of the forehead; the sight was nearly gone." As against the suggestion that the iritis, like the face-ache, might have been rheumatic, he advances the more intense character and more rapid course of that form of the disease.

While, then, we want more evidence, and I must still sum up as against the homœopathicity of Mercury to syphilitic iritis, I think that the case is strengthened in favour of its having power to cause this inflammation in some shape; and I should think the rheumatic variety of iritis — i.e. that which occurs from cold and damp, or in connexion with gonorrhœal rheumatism — the best suited to it, and the corrosive sublimate the best form in which it can be given.

I have dwelt somewhat long on this point, but it is one of importance, and it illustrates the mode of analytical and critical study of drug-action which you may often find it advantageous to follow. Merely saying, in passing, that in what may be called "periosteal rheumatism" — such face-ache, e.g., as results from draughts of cold and damp air, Mercury — again best as corrosive sublimate — is very effective, I pass on to its analogues, which are Kali bichromicum and Phytolacca.

1. Of Dr. Drysdale's provings and presentation of *Kali bichromicum* I do not like to say too much, as Dr. Samuel Jones — who is a humourist — says that it is the only egg of the kind we British have ever laid, and that we never cease to cackle over it. Well: it is not quite the only one, and it *is* a good one; but I mention it now (not to cackle, but) just to say that a revised edition of it is on the point of appearing (with the Aconite and Crotalus of which I have already spoken), and that, among other things, it will shew a striking confirmation — from an old-school source — of the recommendation of the drug in syphilis which Dr. Drysdale made on the strength of its pathogenesis in 1852. It is just in the tissues liable to be affected by both rheumatism and syphilis — the periosteum and the iris — that the chief part of Kali bichromicum as an anti-rheumatic is played. On the periosteum it really exerts a very marked influence, — manifested not only by pain at certain spots in the membrane, but by the characteristic hard swellings of sub-periosteal effusion. These were seen in the provers on the parietal and maxillary bones, and on the tibia (§§ 136, 499, 1450 of Allen). Marked tearing pains were also experienced, especially about the joints, pointing to a similar affection of the other fibrous tissues; and one of the Austrian provers, who took the second trituration, had "several bright red spots and streaks on the white of the left eye," which looks like sclerotic injection.

Kali bichromicum vies (to say the least) with Mer-

curius in rheumatic ophthalmia : it is especially suitable to the form known as "catarrho · rheumatic,"
where the conjunctiva corneæ is involved. Its action
on fibrous tissue has led to its successful use in a
number of local rheumatisms, of the chronic and
"cold" variety. In Dr. Drysdale's arrangement of
it you will find cases of lumbago and sciatica, of
rheumatic headache, and of periostitis, which have
been very satisfactorily cured by it. In the first-
named, as with Rhus, I should think the lumbar
fascia and the sheath of the nerve the seat of the
mischief.

2. It was in affections of these parts that the
power of *Phytolacca* as an anti-rheumatic first appeared ; and a recent case of overdosing by it has
shewn a power on its part of causing periostitis of
the face and forehead. The accompanying symptoms in this case suggested syphilis rather than
rheumatism ; so here again we have a drug belonging to the borderland between these two diseases.
Its differential indications have not yet been established.

V. The power of *Arsenic* to inflame the serous
membranes might have been anticipated to extend to
their synovial analogues ; and the conclusion has
been substantiated by facts. Dr. Imbert Gourbeyre,
who has made Arsenic an almost life-long study, and
to whom we owe so much knowledge regarding it,
has studied this branch of its action in his " Suites
de l'Empoisonnement Arsenical" lately (1880) published in *L'Art Médical.* The articular affections of

the drug are seen chiefly during convalescence from acute poisoning by it.   They chiefly consist of pains ; but in several cases swelling has been observed, and in one at least there was " an intense arthritis of all the large joints."   Dr. Gourbeyre has himself seen a general articular rheumatism supervene on the administration of Fowler's solution for eczema ; and that it was due to the medicine appeared from its recurrence, a year after, on similar treatment being attempted.

Such facts are not unfamiliar to the ordinary works on toxicology, but they have not prevented old-school physicians from treating chronic articular affections with Arsenic.   Phillips cites Haygarth, Bardsley, Christison, Begbie and Fuller (to whom I may add Elliotson) as warranting its efficacy ; and adds — " I quite agree that the remedy promises well in cases where the vital powers are diminished, and the ends of the bones, the periosteum, capsules, and ligaments are swollen ; under the continued use of the drug I have known the joints return to their natural size."   The source of this experience would lead us, in endeavouring to reproduce it, to use the drug in the form of the liquor arsenicalis.

1. I have placed *Acidum lacticum* under Arsenic, because I cannot think that its remarkable power of inflaming the joints — as observed by Foster — should remain without utilisation.   It cannot be employed in rheumatic fever, because there in all probability it is the *materies morbi ;* but in some sub-acute local inflammations of joints, from rheu-

matic causes, it should prove very effective. The articulations affected in his cases were, I should say, those of the knees, elbows, wrists and hands; and they were red, hot, and swollen, as well as painful.

VI. Of *Sulphur*, which as an anti-rheumatic stands by itself, I have only to remind you of what I have written about it in my *Pharmacodynamics*. The reputation it enjoys among the common people for their "rheumatics," carried in their pockets, put under their pillows, or dusted within their stockings, is echoed more scientifically but not more convincingly by the experience gained at sulphureous springs, such as those of the Pyrenees. Its homœopathicity is suggested by the minute dosage thus involved; and is confirmed by the Austrian provings, where rheumatoid pains were very frequent, and one experi-menter, being sceptical as to its influence, tested it by alternately omitting and resuming the drug, when he found his pains invariably shewing a corre-sponding decline or increase. At the same time you will notice that the dosage, though minute, is hardly infinitesimal. Sulphur is not an "anti-psoric" here, and acts better (I think) in the second trituration originally recommended by Hahnemann than in his later 30th.

Such, ladies and gentlemen, is our treasury of anti-rheumatics, and I think we may feel rich alike in their number, in the precision of their indications, and in their tried worth. By our comparative sur-vey of these and of the anti-pyretics we have, I hope, gained some positive knowledge; while we have also learned lessons as to one mode of studying Materia Medica. Our week has been occupied with preparation for two of the oldest and commonest tasks of practical medicine, — the subdual of fever, the relief of rheumatism. "To-morrow" — or rather on Monday —

"To fresh woods and pastures new."

IX.

CEREBRAL LOCALISATION AND DRUG ACTION.

THE field into which I have now to ask you to
follow me is somewhat different from that which we
have hitherto been surveying.   In fever and rheumatism we have taken forms of disease as the sphere of
action of groups of drugs ; and the light which has
illuminated the phenomena we have studied has been
that of pathology.   It is now rather an anatomical
region which will constitute our base of unity ; and,
though pathology will play her part farther on, the
lamp we shall take with us all the way will be that
which is lighted by physiology.   We have to view
the last and crowning discovery of this science in
regard to the functions of the nervous system, and
to see how far our pharmacology can fit in with it,
and make it fruitful for therapeutic ends.

I say that cerebral localisation is the crown of a
long series of researches into the functions of the
nervous system, and localisation has been the key-
note of all previous attainment in this field.   The
first rough survey divided the whole into brain, spinal
cord, and nerves, of which the second was consid-

ered simply a bundle of nerves proceeding from the first, and emerging as the third to be channels along which the "animal spirits," generated in the brain, passed into the frame. The first step in differentiation was to recognise the spinal cord as an independent centre : this was done, from physiological and pathological considerations, by Whytt, Unzer, and Prochaska in the last century, but finally established on anatomical grounds by Gall at the beginning of the present. Then came the epoch-making discovery of Sir Charles Bell (1811), that the anterior and posterior roots of the spinal nerves subserved different purposes, the former being exclusively motor and efferent, the latter sensitive and afferent. As the situation of these roots corresponded, roughly, to a division of the cord into anterior and posterior columns, this was made ; and the former were considered as conveying motor impulses from the brain, the latter sensory impressions to it. Then, when Marshall Hall's researches and writings had revived the notion of the reflex function of the cord, first broached by the physiologists of the last century whose names I have mentioned, its central gray matter was supposed to be the seat of such autonomy as it enjoyed. The brain — the contents of the cranium — next received partition. As the posterior columns of the cord, through the restiform bodies of the medulla oblongata, terminated in the cerebellum, this seemed to be the sensory centre; while the cerebrum was that in which ideas and emotions issued in volitions, which passed through the crura cerebri, the anterior pyramids and corresponding columns of the cord through the motor nerves to the muscles.

So, speaking generally, things were understood during the first decades of the present century. The last fifty years, however, have seen a rapid and steady progress, which, while substantiating some of these notions, has revolutionised others, and has prepared the way for the final conception which it is my present object to elucidate.

The first thing done was the disengagement of the cerebellum and its spinal connexions from all share in sensation and voluntary motion. Flourens found by experiment that its gradual destruction by slicing had no effect on sensibility or on the power of the limbs, but that the faculty of harmonious movement, such as that required for standing, walking, leaping, and so forth, was lost. He accordingly propounded the view that the cerebellum was the organ of co-ordination of muscular motion. Todd, following him, maintained that the posterior columns and restiform bodies were not channels of sensory impressions, but the medium whereby the cerebellar influence was conveyed to the parts specially needing it — the lower extremities. In support of this view he pointed out that the columns in question were uniform in size throughout the cord until its termination in the lumbar enlargement and cauda equina; and that disease in any part of their course caused just such derangement of harmonised motion as the hypothesis required.

Now the facts adduced by Flourens and Todd have
never been disputed; but their theory, though reign-
ing for a time, has always been beset with difficulties.
The phenomena of disease of the cerebellum have
not lent themselves to it : section of the posterior
columns, through which alone the cerebellum can
communicate with the lower limbs, has been found
to produce no impairment in their motorial power,
simple or harmonised ; and, lastly, observations which
I shall hereafter bring before you have shewn that
the path of nervous influence along these columns
and their restiform continuations is not from above
downwards, as the theory would require, but the reverse — not "peripherad," as the electricians say,
but "centrad." Another hypothesis was required
which should embrace these facts as well as those
previously ascertained ; and it has been supplied by
Dr. Ferrier. He maintains the cerebellum to be the
centre of equilibration, the seat of association of
those combined impressions and movements which
result in the preservation of balance under all circumstances. The impressions which convey the
sense of need of adjustment come through three
channels — the eye and optic tracts, the semicircular canals of the internal ear with their afferent
nerves, and the paths of tactile sensibility. In these
last the posterior columns find their place, and the
old conception of their function is somewhat reha-
bilitated. For, though it is not along them that
peripheral impressions travel to become conscious
sensations, they do transmit from below upwards those

tactile changes in the soles of the feet during standing or walking which have so much to do with our
due balance. This is well seen in the disease called
locomotor ataxy, in which they are extensively affected with sclerosis ; where walking is effected by
a series of jerks, and where, if the optical aid to
equilibrium is removed by closing the eyes, the patient immediately feels as if he would fall and has to
grasp at some support. The effect of these visual
impressions ordinarily only comes thus in aid of the
others ; but it is strikingly seen when perverted by paralysis of some of the ocular muscles, whether in disease or from the action of certain drugs, as Gelsemium
and Conium. The part played in equilibration by the
semicircular canals was ascertained experimentally
by Flourens, and has been confirmed pathologically by
the phenomena of Menière's disease. It is through
the fibres of the auditory nerve, of course, that their
influence is conveyed ; but in its perversion there is
no necessary impairment of hearing, as there is none
of vision in the corresponding ocular derangement.
Well : the cerebellum being connected with all these
central paths, and receiving their impressions, trans-
mits them peripherad through the motor paths, with
which it is equally connected by means of its middle
peduncles. Here they are transformed into such
movements as are requisite to preserve or restore the
balance. Through its superior peduncles, moreover,
the cerebellum is connected with the cerebrum, the
seat of consciousness ; and here loss of balance finds
its subjective complement in vertigo, the sense of
giddiness.

I have only been able to give you a brief outline
of these interesting facts, of which however you
have probably heard something from your teachers.
But I would strongly advise you to study the fourth
and sixth chapters of Dr. Ferrier's "Functions of
the Brain," in which you will find the whole matter
fully discussed. His conclusion is that "the cerebellum would seem to be a complex arrangement of
individually differentiated centres, which in associated action regulate the various muscular adjustments
necessary to maintain equilibrium of the body; each
tendency to the displacement of the equilibrium
round a horizontal, vertical or intermediate axis, acting as a stimulus to the special centre which calls
into play the antagonistic or compensatory action."
Accepting this, we may put out of view the cerebellum and its spinal connexions in proceeding upwards to our goal — the cerebral convolutions.

The next great advance in spinal localisation was
made by one in whom America has at least half a
share — Dr. Brown-Séquard. By his experiments
on the cord, and observations of its diseases and
injuries, he has proved that the conductors of sensitive impressions to the brain run in the gray matter,
while — as was always believed — those which convey motor impulses from it occupy (only to a limited
extent, however, as we shall see) the antero-lateral
white columns. He has also ascertained, moreover,
that while the motor tracts cross over to the opposite side — as you know — in the medulla oblongata,
the sensory fibres make their decussation almost im-

mediately upon their entrance into the gray matter.
Disease or injury, therefore, limited to one lateral
half of the cord, will cause paralysis of the same
side of the body, but anæsthesia of the opposite.
Above the medulla oblongata the cross-action is
complete, so that each crus cerebri which emerges
from the pons Varolii represents the motor and sensory functions of the opposite half of the body.

And here again we have to localise. The crura
cerebri are separable into two divisions — the upper
portion, or tegmentum, and the lower, or pes. In
the upper, it is ascertained, run the sensory fibres;
in the lower, the motor. The tegmentum has at its
termination the ganglionic mass called the optic
thalamus, the pes has the corresponding corpus
striatum, which thus, like capitals of columns, crown
the sensory and motor tracts of the cranio-spinal
axis. Being closely connected, they form the double
key-stone of an arch along which impressions may
be converted into movements without the interven-
tion of the higher centres. They stand thus in the
second of the five circuits which such currents can
traverse. The lowest is constituted by those actions specially called "reflex" which have the spinal
cord for their centre, and which are seen even when
this is cut off from all communication from above.
The next appears in the automatic functions of the
medulla oblongata — sucking, swallowing, respiration &c. The third belongs to the mesencephalon
and cerebellum : it is illustrated by the mechanism
of equilibration, as it has just come before us. The

second, as I have said, we have here ; and the first
and highest is that channel for actions requiring
conscious discrimination and voluntary effort which
we are about to consider.

For the basal ganglia — the corpora striata and
thalami optici — are not, as you know, the topmost
of the nervous centres. From them again radiates
a blended sheaf of white nerve-fibres, that terminate
in the cells constituting the gray matter of the cerebral hemispheres, which underlies the whole vault of
the cranium from forehead to occiput. It has always
been recognised that this is the seat of consciousness, the middle point at which mind touches matter ;
where impressions become sensations and the will
developes itself in action ; and in the force generated
wherein alone (in the present state of being) the intellect finds its means of operation. Nothing that
has since been ascertained alters this view of the
position of the cerebrum proper. The phenomena
which some have supposed to indicate a sort of consciousness on the part of the spinal cord, even those
which have led Carpenter to designate the automatic
actions of the basal ganglia as *sensori*-motor, have
been shewn to require no such assumptions. That
there may be consciousness — the subjective side of
impressions and actions — the integrity and co-operation of the gray matter of the hemispheres is indispensable.

In respect of the mode of its co-operation, however, it had hitherto been conceived that the brain
acted as a whole. The sensory nerves coming from

all parts of the frame were integrated in the optic thalamus, and there — through its connexions with the cerebrum — the mind as one perceived them, and translated them into ideas. The influence of such ideation, or the direct commands of volition, were in like manner conveyed to the corpus striatum, and thence to the motor nerves of the muscles. When enquiry was made as to how sensations were discriminated by the mind, or complex actions executed at will, the answer was given by pointing to the continuously separate fibres of sensory transmission, and to the associated motor cells of the anterior cornua of the cord. The former are telegraphic wires, each conveying without confusion its own message: the latter are so many batteries of Leyden jars connected one with another, the electric force distributed uniformly over all the united jars, and all being at the same time uniformly discharged, for which purpose one conductor alone is necessary.

It is sometimes claimed for the phrenologists that they anticipated that differentiation of cerebral function which is now acknowledged. But the conception of Gall and Spurzheim was of a very different order. They divided the mind itself into "faculties," to each of which they assigned a certain portion (of pretty uniform size) of the surface of the brain, ascertaining this by the corresponding elevations or depressions of the cranium supposed to exist in persons variously endowed. Phrenological science, then (such as it was), was a compound of psychology (a very crude one) and craniology; while the art (such as *it* was) should rather have been called cranioscopy. But Gall, who was a man of genius, and who first taught us to examine the brain in a rational way, did really perceive something which lies at the bottom of our present views. The convolutions of the cerebrum, which until quite recently were regarded as only so many folds designed to give ampler surface in

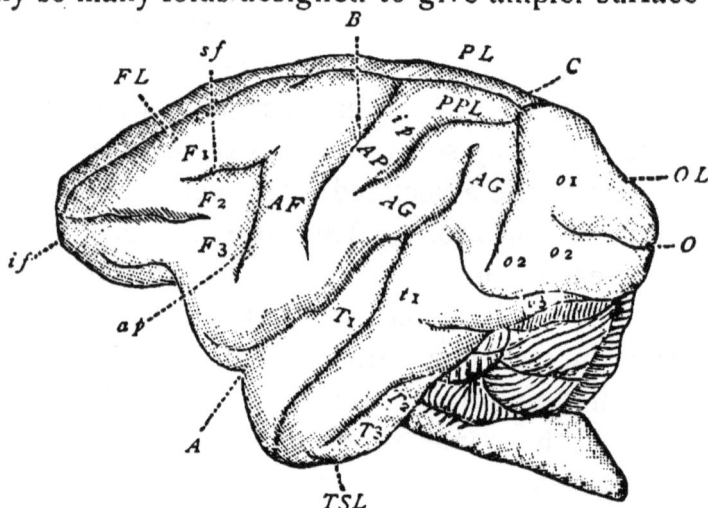

LEFT HEMISPHERE OF THE BRAIN OF THE MONKEY (Macacque).

A. The fissure of Sylvius.   *B.* The fissure of Rolando.   *C.* The parieto-occipital fissure.   *FL.* The frontal lobe.   *PL.* The parietal lobe.   *OL.* The occipital lobe.   *TSL.* The temporo-sphenoidal lobe. — *F1* The superior frontal convolution.   *F2.* The middle frontal convolution.   *F3* The inferior frontal convolution. *sf.* The supero-frontal sulcus.   *if.* The infero-frontal sulcus.   *ap.* The antero-parietal sulcus.   *AF.* The ascending frontal convolution.   *AP.* The ascending parietal convolution.   *PPL.* The postero-parietal lobule.   *AG.* The angular gyrus.   *ip.* The intra-parietal sulcus.   *T1, T2, T3.* The superior, middle, and inferior temporo-sphenoidal convolutions.   *t1, t2.* The superior and inferior temporo-sphenoidal sulci.   *o1, o2, o3.* The superior, middle, and inferior occipital convolutions.   *o1, o2.* The first and second occipital fissures.

a limited space, must — he argued — have a more definite significance, from the regularity with which they occur. So evident was this regularity that physiologists — among whom Gratiolet and Leuret are worthy of special mention — have named them, and the result is what you have all learnt in your anatomy classes.

The two great fissures — that of Rolando above, that of Sylvius below — if caused to meet would about divide the lateral surface of each hemisphere into two equal parts. In front of this line is a vertical convolution, called the *ascending frontal*, and three horizontal ones — the *superior*, *middle*, and *inferior frontal*. Behind it is a corresponding perpen-

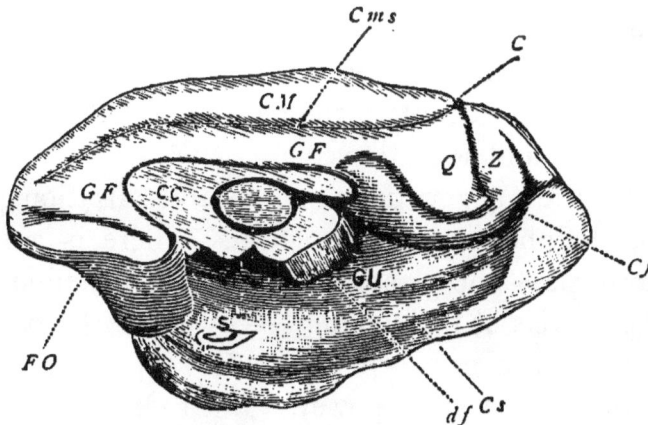

THE INTERNAL ASPECT OF THE RIGHT HEMISPHERE OF THE MONKEY (Macacque.)

C C. The corpus callosum divided. *C.* The internal parieto-occipital fissure. *C m s.* The calloso-marginal fissure. *Cf.* The calcarine fissure. *df.* The dentate fissure. *Cs.* The collateral fissure. *G F.* The gyrus fornicatus. *C M.* The marginal convolution. *G U.* The uncinate convolution. *S.* The crotchet, or subiculum cornu Ammonis. *Q* The quadrilateral lobule, or praecuneus. *Z.* The cuneus. *F O.* The orbital lobule.

dicular fold — the *ascending parietal*, ending above in the *postero-parietal*, and flanked by the *angular gyrus;* while below these, reaching to the base of the hemisphere, are the three *temporo-sphenoidal convolutions — superior*, *middle* and *inferior.* These are all of which — on this surface of the brain — I need remind you; though you will not fail to observe that beyond their range, both in frontal and occipital regions, there is a projecting portion of brain-sub-

stance. But I must ask you also to follow me while
I draw as it were the hemispheres asunder, divide
the corpus callosum, and expose the inner surface.
Here, too, are convolutions, but I need trouble you
with one alone, lying at the base, viz. : the uncinate,
which immediately overlaps the hippocampus major,
and its recurved termination, the "crochet," or subiculum cornu Ammonis.   These are the only parts
of the brain which have as yet been differentiated,
and they seem to contain the centres of all the bodily
movements and sensations.

The earliest suggestion of cerebral localisation grew
out of the phenomena of *aphasia*.   There is a form
of paralysis — the glosso-laryngeal of Duchenne, the
old " paralysis of the tongue " — in which the patient
cannot speak because the muscular apparatus of articulation has lost its power.   But there are circumstances of a widely-different nature under which the
faculty of speech may be lost.   The tongue and lips
may be freely movable.   The integrity of the larynx
appears from the utterance in a natural voice of one
or more words.   But with language in general the
sufferer has as little power of dealing as a deaf-mute.
He does not necessarily lack ideas, but he has forgotten those symbols of our ideas which we call
words.   If you shew him a spoon, he recognises it,
and knows its use.   But ask him if he knows its
name, and he will shake his head.   Ask him if it is a
fork, and he will dissent ; suggest the true name
" spoon," and he will signify approval.   The memory
is therefore capable of being aroused.   But the faculty

of remembering is permanently impaired ; for repeat
your question as to his knowledge of the name a few
minutes later, and he will again sign a negative.

Nor is it the memory of words only that is lost :
the power of articulating them is also absent. Although you have reminded the patient of the name of
the object before him, he cannot pronounce it after
you. He can perhaps say "yes " or " no," though
he often misplaces them : or he utters in reply to
every question, and spontaneously also, some un-
meaning word, as (in two patients of Trousseau's)
"sapon," "cou-si-si." But no conversation can be
held with him.

This is the typical aphasic, though there are great
varieties in degree in particular instances. Refer-
ring you for these, and for a full discussion of the
subject, to those fascinating *Clinical Lectures* of
Trousseau which every practitioner ought to read, I
would say that while aphasia sometimes appears
alone, it is more commonly associated with more or
less hemiplegia. Now, when this latter occurs, it is
almost invariably on the right side of the body, which
implies — as you know — that the lesion is on the left
side of the brain. There, then, for some reason or
other, must be the seat of the faculty of speech.
Now the craniology of the school of Gall had placed
the organ of language in the anterior part of the
brain. In the left frontal lobe, accordingly, the
lesion of aphasia was sought by Dr. Paul Broca, and
was found by him in the posterior extremity of the
inferior frontal convolution on the left side. So

many observations have since confirmed him, and the few apparent exceptions have been so satisfactorily explained, that the localisation has been universally accepted, and the fold of brain is known as " Broca's convolution." The occasional limitation of the mischief is explicable by the spot being supplied by one of the secondary branches of the middle cerebral artery — plugging of which by embolism or thrombosis will cause the symptoms. But why the faculty of speech should be located at this spot, and why on one side only of a bilateral organ like the brain, remained a puzzle as long as the phenomenon stood alone.

In 1870, however, the step was taken which opened the whole field of discovery, and has brought us our present acquisitions. The most fruitful plan by which nervous function has hitherto been ascertained has been the excitation of the part to be questioned by electricity. Up to the time I have mentioned, the few attempts made to electrise the cerebral surface had failed, and it was supposed to be insensible to such stimulus, — as it is known to be to any of a painful character. Now, however, two German physiologists, Fritsch and Hitzig, "established, by a series of valuable experiments on dogs, that the direct application of the galvanic current to the surface of the hemispheres in certain regions caused movements ; and also the more important fact, that definite muscular contractions were associated with irritation of certain circumscribed areas " (Ferrier). They used a continuous current, availing

themselves of the stimulus which follows the making
and breaking of the circuit. In 1872, Dr. Ferrier —
another ornament of my *alma mater*, King's College,
London — began a series of experiments of the
same kind in which he employed the induced current,
thereby getting more prolonged effects, and (after a
time) using monkeys as his subjects and so having
the closest approximation to the human brain. His
results, which entirely agreed with those of Fritsch
and Hitzig as far as theirs went, were published from
time to time in various places, and at last collected
(in 1876) in the volume I have already mentioned.
At first there were many to dispute the validity of
his conclusions, — it being maintained that the phenomena arose from conduction of the currents to
the basal ganglia. This, however, was readily dis-
proved ; and, since then, normal and pathological
anatomy, clinical medicine and surgery, and even
embryology, have come to his aid ; and the doctrine
as a whole must be considered established, whatever
may be the modification in detail it is destined to
undergo.

What, then, is the doctrine ? It is that this middle
region of the cerebral hemispheres, lying about the
fissures of Rolando and Sylvius, embracing the convolutions I have named to you, and corresponding
roughly to the parietal and squamoso-temporal regions
of the skull, is the ultimate seat of sensation and the
starting-point of volitional impulses. The doctrine
further states, that the several convolutions of this
region are distinct centres, motor or sensory ; per-

sistent — so far as they can be followed — throughout
the animal kingdom; and each, in the case of the
motor centres, presiding over definite groups of
movements. Let me tell you what these centres
are, and how they have been identified.

1. Let us begin with the hindermost convolution
— the angular gyrus. Electrical stimulation here
causes movements of the eyeballs, sometimes of the
head, and very often contraction of the pupils. It
might be thought that it was a motor centre for the
eyes. But counter-experiments, in which this area
of the brain was destroyed by the actual cautery,
shewed that the movements were only reflex, excited
by subjective visual sensations (such as occur when
the retina is galvanised). The eye on the opposite
side was not paralysed; but it was blind. Subsequent researches have shewn that this blindness is
only temporary, as long as the occipital lobes remain
intact. Such ample provision is even made for this
important sense, that as long as one occipital lobe
or one angular gyrus remains, vision persists to some
degree in both eyes. It is only when all four are
entirely destroyed that total and permanent blindness results.

Some interesting facts relating to hemiopia and
central vision have come out in the course of this
enquiry, and may perhaps — if time will allow —
come before us. At present it must suffice to
remember that the angular gyrus and the neighbouring occipital lobe — corresponding roughly to the
parietal eminence in the skull — are the cortical
visual centres.

2. On irritation of the superior temporo-sphenoidal convolution in monkeys there is sudden retraction or pricking up of the opposite ear, wide opening of the eyes, dilatation of the pupils, and turning of the head and eyes to the opposite side. These are just such phenomena as would occur when a loud sound was made in the ear, as indeed was ascertained by trial. In the jackal and the rabbit electric stimulation of the homologous spot caused not only the pricking of the ear, but "the quick start or bound as to escape from danger, such as might be indicated by loud or unusual sounds." On the other hand, its destruction by the cautery caused deafness,—on the opposite side if made unilaterally, complete if bilateral. Here, then, we have the auditory centre, at the junction of the parietal and squamoso-temporal regions of the cranium.

3. For the sense of smell it was natural to look to the subiculum cornu Ammonis, as the olfactory tract is very obviously in the lower animals and presumably in man connected anatomically therewith. As a matter of fact, irritation at this spot was attended by a peculiar torsion of the lip and partial closure of the nostril on the same side (for the olfactory tracts do not decussate), such as a powerful or disagreeable odour would excite. Conversely, localised destruction here caused diminution or loss of smell.

4. For similar reasons, the lower part of the middle temporo-sphenoidal convolution, which is in close connexion with the subiculum, has been identified as the seat of taste. The situation of these centres

so near the base of the skull throws light upon the
loss of taste and smell sometimes following blows on
the vertex or occiput, where — as you know — the
chief damage is often done by *contre-coup*. That of
the olfactory centre, moreover, would explain the
occasional occurrence of anosmia in connexion with
aphasia and right hemiplegia, where the lesion must
be close at hand ; and here, as it should be, the anosmia is on the opposite side to the paralysis.

5. Last, the sense of touch has been found situate
in what Dr. Ferrier calls the "hippocampal region,"
including the hippocampus major and the uncinate
convolution, which it is impossible to separate. It
is a special satisfaction to have so traced the ultimate
seat of tactile impressions to the cortical surface,
for it harmonises with the numerous observations
lately made in France on hemi-anæsthesia of cerebral
origin, which has been shewn to result from lesions
independent of the optic thalami, and to be therefore explicable only on the supposition of a higher
centre.

The cerebral seat of all our senses has now been
made manifest : the "five gateways of knowledge"
lead to this tract as the throne-room of its palace.
As in the spinal cord, in the crura cerebri, in the
basal ganglia, the sensory lies behind the motor
region : it is as it were the feminine side of the
nervous system, impressible, perceptive, impulsive,
keeping in the rear of the stronger actor, whose
motions nevertheless it plays no small part in determining. I pass now to this motor area of the hemi-
spherical substance.

The motor centres lie about the fissure of Rolando, being situate chiefly in the ascending parietal and frontal convolutions, but including also the postero-parietal and the posterior half of the three frontals. Thus — proceeding from above downwards — electrisation of the postero-parietal lobule causes (in monkeys) advance of the opposite hind limb, as in walking; of the adjoining upper portions of the ascending parietal and frontal, complex movements of thigh, leg, and foot with — if the electrodes are pushed a little forward — movements of the tail; a little lower, it induces retraction with adduction of the opposite arm; on the ascending frontal, at its junction with the superior frontal, extension forward of arm and hand; all along the ascending parietal, prehensile movements of the fingers and wrist; while, down the ascending frontal, there are centres for supination and flexion of the forearm, for the action of the zygomatics by which the angle of the mouth is retracted and elevated, and for elevation of the ala nasi and upper lip. At the inferior extremity of this convolution, on a level with the posterior termination of the third frontal convolution (Broca's), stimulation produces alternate opening and closure of the mouth, with protrusion and retraction of the tongue. These last movements are bilateral, all the others having occurred only on the opposite half of the body. When, lastly, the posterior half of the superior and middle frontal convolutions are excited, the eyes open widely, the pupils dilate, and head and eyes turn towards the opposite side.

For a counter-experiment, let me read you this from Professor Ferrier. "The right hemisphere of a monkey had been exposed and subjected to experimentation with electrical irritation. The part exposed included the ascending parietal, ascending frontal, and posterior extremities of the frontal convolutions. The animal was allowed to recover, for the purpose of watching the effects of exposure of the brain. Next day the animal was found perfectly well. Towards the close of the day following, on which there were signs of inflammatory irritation and suppuration, it began to suffer from choreic spasms of the left angle of the mouth and left arm, which recurred repeatedly, and rapidly assumed an epileptiform character, affecting the whole of the left side of the body. Next day left hemiplegia had become established, the angle of the mouth drawn to the right, the left cheek-pouch flaccid and distended with food, which had accumulated outside the dental arch ; there being almost total paralysis of the left arm, and partial paralysis of the left leg. On the day following the paralysis of motion was complete over the whole of the left side, and continued so till death, nine days subsequently. Tactile sensation, as well as sight, hearing, smell and taste, were retained. On post-mortem examination it was found that the exposed convolutions were completely softened, but beyond this the rest of the hemispheres and the basal ganglia were free from organic injury." Similar results followed destructive experiments localised in the special centres ascertained to be such by electrisation.

The motor cortical area — the ultimate seat of volition — has thus been identified : it lies, as throughout
the nervous system, anterior to the sensory tract.
Like that, moreover, it is found to be multiple — to
consist of a number of separate centres, each presiding over a definite movement or group of movements.
How, then, are we to conceive that our will operates?
Has it the faculty of singling out the group of cortical cells of gray matter through which it must act if
it desires to execute a given action? We are conscious, in volition, of no such choice ; nor is it easily
conceivable.

Let us go back to our original instance of localisation — the power of speech, and its dependence on
the integrity of the posterior extremity of Broca's
convolution — the third or inferior frontal. What
have our experiments in electrisation of the cortex
taught us about it? Why, this: that stimulation of its
immediate neighbourhood excites the movements of
articulation. Now, if we will think of it, we learn to
articulate before we begin to attach any meaning to
words. The baby is taught to say "papa" first of
all, and then to connect the sound with a gigantic
being in coat and trousers, whom it individualises so
little that for some time to come it will address every
male adult by the name. By degrees, as its education proceeds, we may conceive two processes as
going on in its sensitive and plastic brain. An
organic nexus is being formed there between sights,
tastes and other sensations and the sounds used to
express the objects arousing them ; and further, each

act of articulation is impressing on the cells which
originate it a material memory — as it were, which —
after a time — is revivable at will or on incitement.
Speech, then, — the power of retaining and recollect-
ing words, and associating them with ideas — is the
subjective side of articulation. Let the centre at
which the movements requisite for the latter have
become organised be damaged, and the faculty itself
is lost: there is no store for recollection to draw upon,
no excitable spot at which the process of retention
can be recommenced. The associations with words
remain, when these are supplied ; for they are registered on the sensory centres, which — like the sensi-
tised plate of the photographer — receive them for
ever, and recall them in ideas and emotions.

And now we can see why it is on the left side of
the brain that the faculty of speech comes to be
seated. It is so because it is the result of the muscular actions of articulation. We have seen these to
be performed bilaterally, though the excitation has
been given to one hemisphere only. Potentially,
therefore, the impulse to speech might have proceeded from either side, and been organised there
accordingly ; but actually, that education of the right
arm which we are taught to practise from our early
days so develops the left half of the brain that the
other lies comparatively idle. We know how difficult
it is, when right hemiplegia has occurred, to get the
left hand to execute the required movements — as in
writing. For those, yet more delicate and complex,
involved in speech the task is so hard as to be imprac-

ticable, — at any rate at first. There are cases on record, however, in which — though the paralysis persisted — the power of speaking gradually returned; and here we may fairly conceive the right speech-centre as taking up the duty.

A crucial test of the soundness of this explanation would be afforded by the behaviour of a man naturally left-handed when affected with hemiplegia, for with him aphasia should accompany it only when on the *left* side. Ferrier can refer to three cases in which this coincidence has been observed. And perhaps the practical inference is that which our recently deceased English novelist — Charles Reade — laboured so much to impress, — that we should all endeavour to be ambidextrous. Besides the advantages he urged, we should, if ever the misfortune of a paralytic stroke came upon us, have an arm and hand already educated to do our necessary work ; and, on whichever side it fell, we should preserve our speech unimpaired.

The will to speak is thus the revival, from some sensory excitement, felt or reawakened, of the activity of the corresponding motor centre, and it thus inevitably performs itself in speech. We do not direct the volition to Broca's convolution, which we should find it hard to do : it is there that it originates, and when we say we " will," we mean our consciousness of such awaking. But are all our actions, you will ask, thus mechanically organised, so that on sensation, emotion or idea some habitual action necessarily follows ? Not so, or we should not be moral beings.   There are, in many parts of the nervous system, bundles of fibres whose function we describe as *inhibitory*.   They are themselves motor in power; but their office is — doubtless through their connexions — to restrain such action as would otherwise result from the influence of their fellows.   Thus the pneumogastric — or rather the spinal accessory fibres which run in it — is the inhibitory nerve of the heart : stimulate it, and you retard the pulse-rate ; divide it, and the organ bounds off like a hound slipped from its leash, and palpitates at breathless speed.   Now of such inhibitory power over our impulses to action we are certainly conscious, and there is a portion of the cerebrum as yet unassigned in which probably the power resides.   The frontal lobes, as you will have already noticed, project beyond the motor as the occipital lobes beyond the sensory area.   Stimulation of neither produces any positive effect, nor does any very appreciable result follow their destruction. So far as can be judged, however, in creatures which have no faculty of expression, the occipital lobes are the seat of visceral sensations ; while the removal of the frontal lobes converted rather " bright " monkeys into apathetic ones, wanting in the faculty of attentive and intelligent observation.   Now ethical enquirers have always perceived, when they have thought out the matter, that our moral responsibility lies in our faculty of exercising *attention*.   Self-control is the result of its use.   In refusing to act upon mere impulse, or upon base motives, we are not unimpelled, without motive.   Ideas and emotions — themselves the product of past sensations — still govern us ; but we have chosen to which of them we will attend — which, therefore, shall influence our actions. There are two ways in which, conceivably, we might do this, — one being by the stimulation of the current of ideas we wish to encourage, the other by the repression of those we would refuse. The latter is the more probable alternative; and if, physiologically, the anterior part of the brain be an inhibitory centre, then, psychologically, it is our organ of attention. It is that by which, in obedience to some higher motive, we check the natural impulses to action. It is the organic seat of that which is the distinctively human in us, and accordingly is far more developed in "hoc animal quem vocamus homo" than in any other. In proportion as its powers are educated, we become men of discriminative judgment and deliberate choice : the merely natural becomes less in us, and the moral more : we

> " move upward, working out the brute,
> And let the ape and tiger die."

It is not without foundation, then, that common observation has connected the development of the frontal lobes, and the expansion of the forehead, with the large growth of the truly human in man — with breadth of intellect and moral strength ; while bulk of the occipital lobes would indicate predominance of the animal. And one great encouragement which cerebral physiology gives us is the indefinite capacities of *education* it shews our brains to possess. Happy they who from childhood up have been trained to be no mere creatures of impulse! But they who from any cause have missed this early nurture can do much for themselves to cultivate the inhibitory faculty, and thereby to conduce to their own happiness and that of all with whom they have to do, — to say nothing of higher ends.

It will not have been amiss, ladies and gentlemen, that this discussion of the psychical should have led us to the confines of the ethical, and should have given us a glimpse into the borderland. I am sure that all your teachers here, much as they desire for you to be masters of knowledge, are yet more earnest in their wish that you should be loyal servants of Duty.

X.

CEREBRAL LOCALISATION AND DRUG-ACTION
(continued).

I PROMISED that, at our next meeting, I would enter into some of the anatomical, pathological, and clinical facts connected with cerebral localisation, and so lead on to its connexion with drug-action. I proceed now to redeem my promise.

1. Let me first remind you of some facts to which, during your studies in anatomy, your attention must have been strongly directed, viz. : the distribution of the arteries of the brain. These are, you will remember, the internal carotid and the vertebral. The internal carotid divides into two main branches — the anterior and the middle cerebral; while the basilar artery, which results from the union of the two vertebrals, becomes double again in the shape of the two posterior cerebrals. Take now this semi-diagrammatic view, which our artist has copied from the *Lectures on Cerebral and Spinal Localisation* by Professor Charcot, of Paris, which have lately been translated for the New Sydenham Society. It exhibits the ramifications of the middle cerebral or

Sylvian artery (so called from its entering at the Sylvian fissure). You will see at a glance that the region supplied by its anterior branches is just that which experiment has marked out as the psycho-

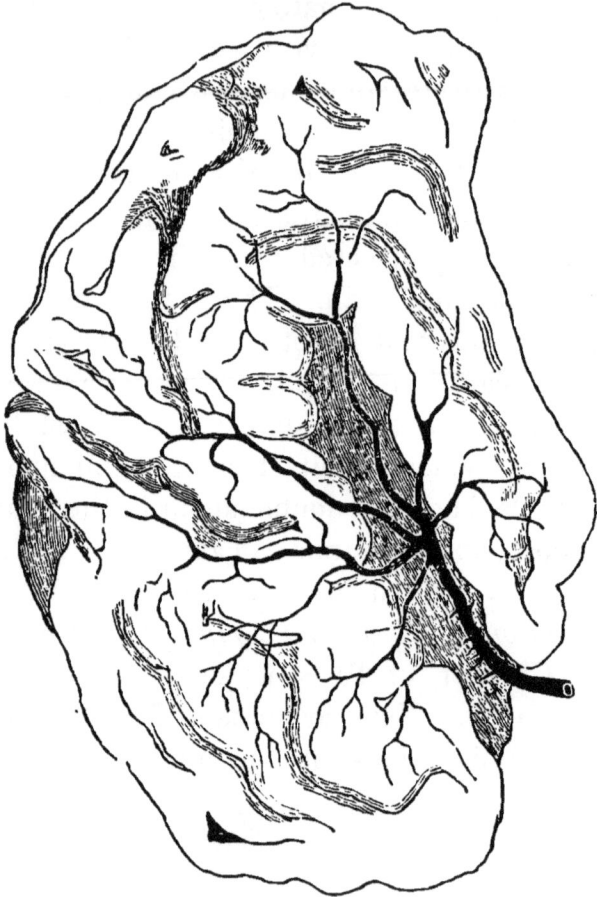

DISTRIBUTION OF THE SYLVIAN ARTERY.

motor area (this is our author's name for the motor portion of the cortex cerebri); that its distribution explains why the posterior portion of the three frontal convolutions should be functionally different from the remainder (which is supplied by the anterior

cerebral) ; and that aphasia may well exist alone,
since Broca's convolution has a branch to itself.
Let this be plugged by an embolism, and atrophy
of the seat of speech may be the sole lesion, sudden aphasia the sole result. This actually occurred
in a case observed by Charcot at the Salpetrière.

The posterior branches of the artery, you will
observe, reach the angular gyrus and the temporo-
sphenoidal convolutions, which have been ascertained to be the seat of the senses of sight and
hearing, which through the circulation therefore —
as well as (doubtless) through innumerable commissural nerve-fibres — are associated with the motor
centres. But remember that the posterior cerebral,
besides its chief work of supplying the occipital lobes,
sends branches also to the temporo-sphenoidal region,
to the gyrus uncinatus, and to the hippocampus.
As you will remember that there is somewhere in
the occipital lobes a supplementary visual centre,
there are some features here suggesting special provision for the integrity of regions which are the seat
of our conscious sensations, and thus of our ideas.

2. Again, I must anticipate morbid anatomy so
far as to say that, when the psycho-motor area of
the brain is seriously affected — as with softening,
a peculiar degeneration — a sclerosis — is found to
descend therefrom through certain definite tracts
of the spinal cord. These are the small columns
called (after the first observer of these degenera-
tions) Turck's, which lie immediately beside the
anterior median fissure, and a certain portion of the

lateral columns. Such tracts, therefore, must be regarded as the true prolongations of the fibres coming off from the motor cerebral cells and proceeding down the crura; the rest of the spinal fibres being (longitudinally) commissural. Now it is an interesting fact that these cerebro-spinal fibres are latest in development, — that their axis-cylinder (which you know is the central and essential part of the nerve-fibre) has not acquired its myeline sheath, and so is not insulated for action, at birth. This is ascertained by treating the cord of the newly-born child with osmic acid, a substance which gives a black stain to the sheath — the white substance of Schwann, as we used to call it, but leaves the axis-cylinder clear. Those parts, I say, which in paralysing cerebral lesions are affected with descending sclerosis, are just the parts which remain clear under the osmic acid. The true spinal cord of the infant is already fully developed, as also his medulla oblongata, for these are the seat of the automatic reflex activities it immediately requires. The fibres which convey impressions to and from consciousness are not yet needed. The cerebrum itself is in the same rudimentary condition, — its main structural details barely outlined, and the nerve-tubes almost universally absent. Moreover, in the young of those animals which are born blind, and which therefore — like the human infant — are at this time mere reflex automata, electrical stimulation of the excitable portions of the cortex determines no movements. It does so in animals which are born with open eyes,

and the histological development and chemical constitution of the cerebrum of these creatures differs
little from the condition which obtains in the adult
state.

3. Anatomy can go even farther still — into the
realm of histology, and still find facts confirmatory
of the doctrine of cerebral localisation. It was long
ago noticed (I mentioned it in an account of the
nervous system which I contributed to the *British
Journal of Homœopathy* in 1861) that the gray cells
of the anterior cornua of the cord — whence, as you
know, the motor roots of the spinal nerves take
their origin — were considerably larger than those in
the posterior cornua, which are sensitive. Now of
late the cells of the gray matter of the brain have
been examined microscopically, and similar differences have been found to obtain. In front of the
fissure of Rolando cells so large as to be called
"giant" are to be found, and they are especially
massed about this sulcus, i.e. in the neighbouring
parts of the ascending parietal and frontal convolutions — the great motor area. Behind this spot, in
the sensory region of the brain, the cells — the feminine cells, you will remember I called them — are
markedly smaller, though still superior to those of
the cerebellar hemispheres.

4. I now come to the bearing which pathological
anatomy has upon our present subject, and in this
part of my task am more than elsewhere indebted
to Professor Charcot's lectures.

If the inferences drawn from physiological experiment are correct, disease — as from softening due to embolism or thrombosis occluding the anterior branches of the middle cerebral artery — affecting the psycho-motor area should cause hemiplegia, even though the corpora striata remained untouched. This corollary has been abundantly warranted by experience. Charcot has recorded all the cases of ischæmic softening of the cortex of the hemispheres dying in his wards in the Salpetrière during the last fifteen years.

" In these cases " he writes " the lesion presented itself under the form of yellow softening (*plaques jaunes*), more or less extensive in area, involving to a variable depth the subjacent white matter and occupying the most diverse regions of the surface of the hemisphere. It is expressly mentioned in all the observations that the softening had not affected the central masses — optic thalami, corpora striata, and internal capsule. My observations may be divided into two groups.

" The first includes the cases in which permanent hemiplegia had not existed during life, and in which secondary degeneration was found at the autopsy to be absent.

" In all, the convolutions supplied by the Sylvian artery, and particularly the ascending frontal and parietal, remained intact. The yellow softening was situated in one of the following regions, viz.; some part of the sphenoidal lobes, the lobulus quadratus, the cuneus, one or both occipital lobes in their entirety, some part of the anterior two-thirds of the frontal lobes.

" In all the cases of the second group there had been, on the other hand, permanent hemiplegia and well-marked secondary sclerosis. The feature common to these cases is that the lesion invariably involved, to a greater or less extent, one or other of the ascending frontal and parietal convolutions, principally in their upper half, and often both at the same time. In addition, the regions nearest to the frontal and parietal convolutions were very frequently involved."

This should be enough. Here is no single case, which might be exceptional and neutralised by others of opposite significance ; but a series of observations in a large special hospital extending over fifteen years, and all concurring — positively or negatively — in the same conclusion. The statement made by Nature in answer to experimental enquiry is volunteered by her of her own accord.

Let me now say something of this " descending sclerosis," which has more than once come before us. It consists, as I mentioned, in an indurative degeneration of those tracts in the spinal cord which are direct prolongations from the brain — the columns of Turck and the outer and posterior border of the lateral columns. The former are not crossed, and do not penetrate very far ; the latter are continuous with the decussating fibres in the anterior pyramids of the medulla oblongata, and are altogether so much more important a factor in the matter that the whole thing is called " lateral sclerosis." This morbid process, in its secondary descending form, will follow any lesion of the motor tract, from the white substance of the hemispheres downwards, which interrupts the continuity of its fibres. It does *not* follow upon hemiplegia consequent on mischief limited to the corpus striatum ; which appears to shew that this body is a ganglion ancillary to motor agency, rather than a secondary centre to which the whole tract converges to radiate forth again.

The explanation of this secondary degeneration is derived from the phenomena which occur when a

nerve is separated from its centre. When the mixed
spinal nerves are divided, the peripheral end degen-
erates, both in its centripetal and its centrifugal
fibres. But when it is — as it were — unravelled by
ascending to its roots, a different result is obtained.
When the anterior root is divided, its peripheral
end is still that to wither. The motor centres of
these roots are thus (as we shall see from other facts)
their trophic centres also, viz. : the cells of the anterior cornua of the cord. But divide the posterior
root — which you must needs do before you get to
the ganglion which you know exists upon it, at its
junction with the anterior ; divide this root at any
point, and it is the central end which withers — the
peripheral remaining intact. The ganglion, then,
whose presence in my student-days anatomists used
to consider somewhat of an impertinence, is now
shewn to be the trophic centre of the sensory root,
— nay, of the whole sensory nerve ; for extirpate it,
and degeneration attacks not only the posterior root,
but all the centripetal fibres of the mixed nerve.

Lesions of continuity of the cerebral motor tract
thus cause degeneration below their seat, because
they cut off the fibres descending therefrom from
their trophic centre, which is the psycho-motor area
of the cortex. What symptoms do they cause?
Two only, invariably : loss of power and contracture
of the limbs. These are usually and mainly seen on
the opposite side of body to the lesion ; but sometimes or to some extent on the same, — the latter
being accounted for by the involvement of the col-

umns of Turck, which — as I have mentioned — are
not crossed. Now this contracture of the limbs is a
phenomenon in every way of great interest. It has
long been known as a symptom of much import in
the prognosis of hemiplegic attacks. "Early rigidity" means little, as it may result merely from irritation, as about a clot; but "late rigidity" — that
which supervenes some months after the "stroke"
— implies that the paralysis is going to be permanent. Dr. Todd ascribed it to cicatrisation of the
brain at the injured spot; but the descending degeneration since discovered is more of a *vera causa*, as
amply accounting for the impossibility of recovery.

The lateral sclerosis, nevertheless, explains only
the paralysis in these cases : but what is the cause of
the contracture? This, it would seem, must be
something of a functional rather than an organic
nature, for this reason, that the rigidity is not uniform and constant. It relaxes somewhat during rest
and sleep; it is increased on movement, and especially by anything like exertion (I mean of course
on the healthy side); its appearance may be anticipated, or its intensity enhanced, by the administration of strychnia, while opposite results follow large
doses of bromide of potassium. Now what these
agents increase and diminish respectively is the reflex excitability of the spinal cord; and, looking
farther in the direction thus suggested, we come
across some very interesting facts. There is a symptom, present in certain kinds of paralysis, absent in
others, to which the Germans have given the name

of "foot-phenomenon" or "ankle-clonus"—the latter being its usual designation in England. It is elicited by supporting (as with the hand under the ham)
the paralysed leg, so that it may hang loose and swing.
Then, if the point of the foot be suddenly raised, "a
series of shakes is at once provoked, which collec-
tively constitute a kind of rhythmical movement or
oscillatory trembling more or less regular and persistent" (Charcot).   Now in the healthy subject
this "spinal trepidation," as the French call it, cannot be produced, and it is absent in such affections
as locomotor ataxy and infantile spinal paralysis ;
but it is present in all paralyses in which contracture
exists and tends to be established.   In hemiplegia of
cerebral origin, it often precedes the "late rigidity"
by some weeks : it is thus of value for prognostic
purposes, and it further throws light upon the nature
of the symptom it heralds.   For it evidently belongs
to the group of phenomena known as "tendon-re-
flexes," of which the starting forward of the leg on
striking the patellar tendon is the most familiar ; and
these, as their name implies, are *reflex* actions.   If
the ankle-clonus, then, absent in health, is present
in paralysis with contracture, it implies that in this
condition the reflex excitability of the cord has
undergone enhancement, i.e. (as there are no sensory
phenomena here) that the cells of the anterior cornua
are in an irritable state.   What should induce them
to be so ?   This question is probably to be answered
by considering the destination of the cerebro-spinal
lateral fibres — the "pyramidal tracts," as Charcot

calls them.   They progressively diminish in size from
above downwards, are " used up " — as it were — in
the course of their descent.   They convey the orders
of the will to the motor nerves, and yet there is de-
monstrably no continuity between them and the anterior roots.[1]   What remains, then, but that they end in
the cells of the anterior cornua, from which the motor
roots emerge?   If they do so, then it would seem
that — in descending sclerosis — these cells must
either take on the process of degeneration, or must
oppose it.   That they do not ordinarily accept it appears from the fact that " the muscles of the extremities on the paralysed side in patients affected with
permanent hemiplegia of cerebral origin present no
other atrophic changes than those resulting very
slowly from the functional inertia to which these
muscles are condemned " (Charcot).   Did the anterior cornual cells suffer, rapid muscular atrophy
would follow, as indeed occurs in the acute disease
known as anterior poliomyelitis (of which infantile
spinal paralysis is an example) and in the chronic one
we call " wasting palsy " — the " progressive muscular
atrophy " of Duchenne.   Sometimes, in ordinary hemiplegia, this "amyotrophy" does occur, and the cells
of the anterior cornua have been found to have suc-
cumbed ;[2] but more commonly, as I say, they resist.
Now resistance implies over-action, irritation.   Given
an irritable state of the cells of the anterior cornua,
and you have the exaggeration of reflex activity
shewn in the ankle-clonus, you have the exaggera-

---

[1] See Charcot, *op. cit.*, p. 299.   [2] Charcot, *op. cit.*, p. 201-3.

tion of the normal muscular tone which appears as contracture.

The point of special value to us here is that such a condition is, as Charcot says, "purely dynamic and corresponds to no appreciable anatomical modification." Hence its variability under circumstances; and hence (a point I have not yet noted) its occasional diminution and even disappearance in the course of time. Now all this brings it into the therapeutic sphere. We have seen that contracture may be anticipated or intensified by strychnia so given as to induce its physiological effects. Hence the warnings of old against using Nux vomica or its alkaloid in cerebral hemiplegia; but hence, obviously, the indication for them to us, in our non-physiological doses. Even in the old school Mr. Charles Hunter finds that by reducing his dose of hypodermic strychnia to the sixtieth or eightieth of a grain he can remove "the muscular twitchings, spasms, or cramps of the paralysed parts;" and these are just the variable element in contracture, and one chief source of its irksomeness. We cannot (probably) by drug-treatment alter the sclerosis, and so cannot restore power; but it will be no slight help to our patients if we can further the natural possibilities of mitigation or subsidence of rigidity. Let me also remind you that if at any time a sudden accession of wasting —with loss of electric contractility—should occur in the paralysed muscles, indicating commencing degeneration of the cells of the anterior cornua, we have a hopeful remedy in Plumbum. Lead-palsy is

the precise analogue in drug-disease of progressive
muscular atrophy; and even in what may be called
the acute form of that affection — in anterior poliomyelitis — the drug would seem effective. Dr. Jousset has recorded — in his *Leçons de Clinique Médicale*,
which Dr. Ludlam has done such good service by
rendering into English — an undoubted case of the
kind, in which Plumbum (in the 30th dilution) effected a cure.[1]

All this about "late rigidity" has been somewhat
of a digression; but I think it warranted by the
frequency with which cerebral hemiplegia will come
before you in practice, and the desirableness of your
knowing all that can be known about it. I return
now to the more direct thread of our discourse; and
proceed to say something of the clinical aspects of
cerebral localisation.

5. If it be so, that certain spots in the cerebral
cortex are the seat of definite functions, sensory or
motor, ought not the knowledge of them to be of aid
to surgery? If symptoms of pressure on the brain
exist after an injury, or of localised irritation where
an exostosis or other morbid growth may fairly be
presumed to exist, would not the surgeon be justified
in trephining the skull at the point indicated by the
functional disorder? This question was largely discussed in Paris during the years immediately following the researches of Fritsch, Hitzig, and Ferrier,

---

[1] Cuprum, I may add, is another medicine which has amyotrophy among
its effects, and also something very like contracture (see my *Therapeutics*, i.
236).

especially in 1877, when the French medical journals
were full of it.   On the whole, the verdict was in
the negative.   In injuries, the old rule of trephining
(if at all) at the seat of the wound seems still to be
the safest; while, in the case of morbid growths,
the pressure they exert is so diffused as to make it
almost impossible to identify their exact seat.   You
know, for instance, how often they cause optic neu-
ritis, without interfering in any obvious way with
the integrity of the optic tracts.   All this is true,
but I think that an exception might be made in the
case of presumed exostosis.   If troublesome symptoms — epileptic or otherwise — were present, and
the history of the case raised a reasonable probability of such a growth being their cause, I would
strongly advise the application of the trephine at the
spot to which the symptoms pointed.   As the dura
mater need not be opened here, the operation would
involve little risk — the usual precautions being
taken; and sometimes a deliverance otherwise impossible might ensue.

6. Of yet more interest, to most of us, are the
phenomena which clinical medicine presents in this
sphere; and they have the further recommendation
that some of them at least suggested the doctrine
of cerebral localisation before physiological experiment had established it.   I refer to the observations
of Dr. Hughlings Jackson, of London.   I wish that
this philosophical physician would do himself the justice and us the benefit of collecting his scattered
contributions on the subject to journals and transac-

tions into one work. Dr. Ferrier spoke of him as
doing so in 1876; but the promised "Clinical and
Pathological Researches on the Nervous System"
have not yet seen the light. Dr. Jackson was struck
with the unilateral and localised forms sometimes
presented by epileptic convulsions. On the theory of
the disease being resident in the medulla oblongata,
the whole frame should be uniformly convulsed. The
limitation often seen to one side of the body disal-
lowed this conception, at least in such cases, and
compelled that of a direct discharge of nerve-force;
while again the involvement only of the arm, or leg,
or certain of the facial muscles prevented reference
to the gray matter of the corpus striatum, lesion of
which influences the whole of one side of the body.
Thought was thus inevitably carried up to the hemispheres, and he drew the inference that the convolutions surrounding the striate bodies had a direct
relation to movements, — the convulsive phenomena
being the result of irritative or "discharging" lesions
of the cortex in this region.

In several places he has explained his views about
these "discharging" lesions. He distinguishes them,
by this name, from "paralysing" lesions: they are
to one another as electrisation of the cortex to destruction of the same spot in physiological experiment. The discharging lesion, like the former,
causes over-action; not better function, but "instability" of the gray matter in its neighbourhood, leading to frequent, involuntary, violent exercises of its
influence. If it be a motor centre, this will result

in localised convulsion; if a sensory one, in some subjective manifestations of its form of sensibility. The first is easy of conception: to illustrate the second, let me read you the narrative of a case communicated by Dr. Gowers to the *Lancet* of 1879.

"The patient" who was a man of about 30 "stated that he was well until two months before he was first seen, when one morning something seemed very brilliant before him, 'as if he had a polished plate on his breast.' He felt giddy, but did not fall; he sat down, bathed his head, and was better, but afterwards felt extreme pain in his eyes, 'as if they were bursting.' Subsequently, he had slight attacks daily, of the following character: A pain commences in the neck, goes across the head, comes down between the eyes, and is felt on each side of the bridge of the nose. If walking, the road or path seems to get narrower and narrower, so that he hardly knows where he is going, and simultaneously his sight fails; he feels with a stick to see if he is not getting off the path. The pain at the top of the head and in the eyes is something awful, and the eyes seem to throb. The loss of sight is not complete; he can only see just before him, nothing on either side; but he can generally see better to the right than to the left. During the three days before he was first seen, at the commencement of the attacks, as the sight was going, he had a flickering of light, 'like a gold serpent,' in the eye, moving in all directions very fast; seen with both eyes, he thought, but more before the left eye than the right.

"He was treated with bromide and belladonna, and the attacks ceased, except that once or twice he had a slight flickering before the eyes. He had, moreover, several attacks of pain on the top of his head, coming on suddenly, and he often had pain at the back of the neck."

After continuing well for five months, this man had a fall, striking the head. General symptoms of cerebral disturbance followed, and he died comatose. At the autopsy, a large tumour was found in the right occipital region, invading the angular gyrus. Now this convolution, you will remember, is the special seat of the sense of sight. The neighbouring tumour acted as a discharging lesion to it and caused instability of its gray matter. Hence the paroxysms, so to speak, of subjective visual sensibility — at first exaggerated, later diminished. Observe, further, the effect of treatment in removing, almost absolutely, this part of the trouble; so that while the tumour remained, it could no longer cause morbid discharge of the gray cells in its vicinity. Does not this remind one closely of lateral sclerosis of the cord and the contracture caused by the irritation of the anterior cornua? Here, as there, the inference is therapeutically encouraging: part of the patient's trouble is functional only, and may be removed by medicines. Of those actually used, the combination spoils the effect for any clinical inferences; and indeed is an unusually strange one. A bromide simply dulls all sensibility, morbid or healthy; but why Belladonna? If this had been

given alone, we should have called the cure a beautiful piece of homœopathy; for no drug causes so
much visual hallucination, and this patient's can be
exactly paralleled from its pathogenesis. One of
Hahnemann's provers, Kummer, who evidently experimented on a very sensitive woman, gives among
her symptoms — "she sees on the ceiling of the
room a white star as large as a plate, and light silver
clouds pass over it from left to right." This occurred several times and in various places. We
should certainly give Belladonna when discharging
lesions affected the visual centres, and might expect
the best results.

I would direct your attention, before leaving this
case, to the seat (or seats) of the pain experienced.
This was evidently connected with the visual excitement, and yet we know that the optic tract itself
has no common sensibility. As this is supplied to
the eyeballs by the fifth nerve, and as it was in them
that the pain was first experienced, we might suppose the commencement of the feeling at the back
of the neck to be connected with the origin of this
nerve at the base of the brain. But why the going
across the head? and why the subsequent sudden
attacks of pain "at the top of the head"? This I
cannot say; and the angular gyrus, you are aware,
corresponds rather to the parietal protuberance.
But I remember a case of brain-fag in which I was
consulted, and where one of the symptoms was great
sensibility to light. The patient said that when he
drew up his chamber-blind in the morning, and the

sun happened to be shining brightly, a sudden pain
struck through his eyes to the top of his head. (I
may mention that this symptom, with the visual sensitiveness in general, disappeared under Nux vomica
30.) Whatever it means, then, I think we may hold
— provisionally at least — that the pain which the
optic tract cannot feel for itself will be referred to
the vertex.

Dr. Hughlings Jackson has described cases of limited paralysis and convulsion as "the results of experiments made by disease on particular parts of the
nervous system of man." A case of this kind, which
displays a sort of sensory convulsion, is such an experiment, and it is full of instruction. It suggests
the meaning of such as those we often encounter,
where one or more epileptiform attacks occur in a
man to whom they have been hitherto unknown, and
where there is no syphilitic history to suggest exostosis, and then are followed by sudden hemiplegia.
Here some lesion, probably a small aneurism in a
cerebral artery, has first by its irritating presence
caused discharge in its neighbourhood; and finally,
giving way, has become paralysing to the same spot.
Again, it throws great light upon the "aura epileptica." You know that epileptics frequently experience, as premonitory to their paroxysm, a sensation of
some kind, —a sight, a sound, a smell, a taste, or the
"aura" strictly speaking — a feeling as of a breath
of cool air along a limb. In all cases, the sensation
becomes more intense as the fit draws on, and — if its
nature allow — will seem to be approaching the brain.

The aura will ascend the limb, the supposed object of sight will advance. A patient of mine used to see a pair of white wings drawing rapidly nearer to her from the other end of the room, and as they reached her she lost consciousness, fell, and became convulsed. In the light of what we have learned it is evident that this means a state of instability of some sensory centre, leading to recurrent discharges, which ultimately affect the motor apparatus, but first shew themselves in the sensations proper to its own endowments, — these being, by a well-known law, referred to the periphery first of all or undergoing a corresponding increase in intensity. Sometimes, indeed, the aura (not then strictly to be so called) is motor from the outset, and consists in twitching — say — of a thumb or finger of the arm which is ultimately the main seat of convulsion.

When these auræ were first studied, the idea of reflex action was dominant, and it was supposed that something could be done by dividing nerves along which they seemed to proceed, or forcibly compressing the limb above them. No constant result followed these attempts, and we see how it must have been so. They were but peripheral expressions of central irritation. If anything was done, it was by indirectly modifying this; and such practice must be very uncertain. We should have a much surer and finer road to the affected part if we possessed drugs which have shewn themselves capable of arousing similar subjective sensations. When any such symptoms have occurred in pathogenesy, we should note them, and may be able to turn them to good account.

For instance, Carbolic acid has shewn a power of enhancing the olfactory sensibility. In two of its provers this occurred from simply smelling at the acid ; it was associated with the frontal headache of the drug, and — when it once came on — lasted for hours. In another it resulted from involuntary exposure to the vapour, and was intense, making the presence of any odorous matter intolerable. In a case I have myself observed, a prolonged inhalation caused all food to smell badly for weeks after. Well : it was perhaps a small matter that this symptom enabled me to remove with the drug a frequently-recurring headache in a lady not subject to such trouble. There was nothing distinctive about the pain ; but, she said, whenever it comes, everything seems to smell too strong. Carbolic acid 12 rapidly and permanently checked the attacks. But, on another occasion, a clergyman consulted me with this story. He had been in fair health till some months previously, when he had incurred measles. On recovering from this, everything with any odour had begun to smell disagreeably to him. This had increased to such a degree that he could not take adequate food from the disgust it caused him : he had lost many pounds in weight, and was getting weak. It was not a case of ozæna, for there was no discharge from the nostrils, and no offensive odour of the patient's breath or nasal mucus. I could only conceive that the morbillous catarrh had in some way affected the olfactory nerves ; and I prescribed Carbolic acid. As he said that he was an old homœopathist, and very sensitive to medicinal action, I gave him the 30th. At the end of a week he reported improvement ; and in three weeks time he had lost all his bad smell, and was enjoying his food like any other man.

Now with this action of Carbolic acid, thus observed and thus verified, if we had a case of epilepsy in which the premonitory symptom was a subjective odour, I think we should be justified in expecting the best results from it.

At our next meeting I shall hope to draw out more fully the possibilities of drug-action in connexion with the localisation of nervous function.

XI.

CEREBRAL LOCALISATION AND DRUG ACTION
(*continued*).

SEVERAL times already, in our study of cerebral localisation in its pathological and clinical aspects, we have touched upon its connexions with drug-action. This evening I propose to concentrate your attention on this branch of the subject. It is that which especially belongs to us of the homœopathic school; and we are disposed to maintain that it is at least not inferior in importance to any other. We frankly acknowledge the industry and acumen with which the subject at large has been worked out by our brethren in the other camp; but, save for prognosis, and for an occasional tentative surgical operation, they must feel that the knowledge gained has done little for therapeutics — that one great end of the physician. Let it be ours to emulate their zeal and to utilise their results by seeking to make them fruitful in this direction. The law of similars enables us to do this. We have acquired a better understanding of a group of morbid phenomena: let us find parallel facts in the physiological action of drugs, and unlimited possibilities open to us in the way of benefit.

I. I have spoken of Hughlings Jackson's "discharging lesions"—of those which cause instability and morbid outbursts of function on the part of the gray matter in their vicinity; and have shewn that this instability, being a purely dynamic disturbance —an irritation, ought to be under medicinal control. We have seen two instances of it,—the twitching, cramps, spasms of paralysed limbs, ultimating in permanent contracture, which arise from the effect on the cells of the anterior cornua of lateral sclerosis of the cord; and the subjective visual disturbances resulting from a tumour adjoining the angular gyrus. In the latter case Belladonna, which would have been our own remedy, was one of the drugs administered with good effect: in the former Strychnia, known to increase the symptoms, and obviously homœopathic to them, has already acquired some repute against them, and in our hands may fairly be expected to accomplish more.

1. Looking farther in the same direction our attention may be drawn to the disease known as locomotor ataxy—the "tabes dorsalis" of Romberg, who was the first to describe it. The clinical history of this spinal affection you have doubtless learned from your teachers, and I will only go over the ground again so far as to remind you of a point in its pathology. It used to be spoken of as a sclerosis of the posterior columns of the cord. But localisation has been at work here also, and has shewn that

under this anatomical name there are comprised at least two distinct tracts of nerve-fibres. The one pair is immediately contiguous to the posterior median fissure: they are called the columns of Goll. The others, known as the columns of Burdach, intervene between those of Goll and the posterior nerve-roots. It is in these last — the columns of Burdach — that the sclerosis which lies at the bottom of the phenomena of locomotor ataxy is primarily seated, and to these it may be limited.

Now it is well known that, as Hammond says,[1] "locomotor ataxia often spontaneously remits in the severity of its symptoms: indeed, the remission may at times amount to a complete intermission." This intermission seems sometimes to have been obtained to so prolonged a stage as to warrant belief in a cure by the internal use of nitrate of silver; and of late has been frequently secured by the process of stretching the sciatic nerves. It seems hardly conceivable that, spontaneously or by such measures of our art, a degeneration like sclerosis can alter for the better, however its progress may be checked. There must therefore be some fluctuating, functional, dynamic, and therefore modifiable element in the disease. Its primary manifestations are (I speak of the ordinary form) in the sensory nerves of the lower extremities, these being the seat of lightning-like flashes of neuralgic pain — the *douleurs fulgurantes* of the French, and of loss of function. By the suspension of reflex activity thus induced the muscular coordination

[1] *Diseases of the Nervous System*, 7th ed., *sub voce.*

necessary for standing and walking is impaired, and hence the gait which gives the name "ataxy" to the disease. These pains and this anæsthesia are ordinarily set down to compression of the fibres of the sensory roots, as they pass through the hardening columns on their way to the gray matter of the cord. But, if so, how could there be any variability in them, any long suspension or permanent removal of them? Is it not far more likely that the trouble lies in the cells of the posterior cornua — the source of the energy of the sensory roots? The columns of Burdach are not made up of long fibres, stretching all the way up the cord, but of short ones, commissural between one segment and another. They thus arise from and end in these very cells ; and as we have seen lateral sclerosis irritating the anterior cornua by a similar connexion, we have a right to invoke the same agency here.

I would submit, then, that in not too advanced cases of locomotor ataxy, the symptoms — at any rate the distressing pains — are due to a dynamic irritation of the gray cells of the posterior cornua, and may be amenable to treatment. Later, the cells submit to the degeneration, and this extends to the roots, and then nothing is to be done ; but I speak of an earlier stage. The medicines I would urge for consideration here are Agaricus, Arsenicum, and Belladonna.

Agaricus developed, in the heroic Austrian provings, a marked action on the spinal cord ; and among the symptoms were fugitive neuralgic pains along

the spinal nerves. Dr. Dyce Brown, in a valuable
study of the pathogenesis which appeared in the
twentieth volume of the *Monthly Homœopathic Review*, suggested its appropriateness to the pains of
ataxy.   I have said that I could not quite follow him
here, as the inflammatory induration which lay at
the basis of these was beyond the range of action
of the drug.   They answered rather, I suggested,
to those of what is called "spinal irritation."   So
far I was right ; but we have seen that the *douleurs
fulgurantes* are just an expression of such irritation,
which intervenes — as it were — between them and
the sclerosis.   Dr. Brown's recommendation was
well-founded ; and Dr. Sauer, at a late meeting of
the Silesian Homœopathic Society, spoke of having
seen at least temporary benefit from the remedy.
You will remember that the neuralgic pains of Agaricus are compared to sharp ice touching the spots, or
cold needles running through the nerves, — in this
contrasted with those of *Arsenic*, where the imaginary
needles are red-hot.

Turning now to the drug last-named, I would say
that the "myelitis" which has been observed from
it is hardly that of locomotor ataxy, so that it has no
fundamental relation to the disease.   The arsenical
paralysis, however, which is generally an epiphe-
nomenon of poisoning by the drug, is always accompanied with neuralgic pains and usually with loss of
sensibility, at least to everything but cold.   A similar
exemption of the sense of temperature is often seen
in the anæsthesia of ataxy, which, too, not uncom-

monly sets in with *burning pains* in the soles of the
feet, and "pins and needles" or other forms of
numbness.

Belladonna I recommend here rather on pathological grounds, for I do not feel sure that the group of
symptoms I have adduced in my *Pharmacodynamics*
will bear much weight as parallels.   Its tactile anæsthesia is certainly not central, like that of the disease, and it has never — save in one passing instance
— caused neuralgic pain.   But the grounds on which
we nevertheless administer it as an anti-neuralgic,
and with brilliant success, hold good here also ; and
I should rely on it actually to preserve the gray cells
from invasion, and so to limit and perhaps starve
out the disease.   I certainly think I once made a
cure with it, in an incipient case, giving the 1st decimal dilution.   ,

2. We have, I think, another example of a discharging lesion in *neuralgia*.   In this instance there
is not, ordinarily, any substantive growth or change
in the neighbourhood — any tumour or sclerosis — to
set up the discharge : the lesion is the quasi-inflammatory state of the gray nucleus of the sensory
nerve in which the pain resides.   I assume that the
work done by the late Dr. Anstie in reference to
neuralgia has definitively proved it to be (save of
course when traumatic) of central origin.   His small
volume on the subject [1] is worth its weight in gold :
it is "hewn from life" and crammed full of thought.
I differed from my old fellow-student at the time of

----

[1] *Neuralgia and the diseases that counterfeit it.*   London, 1871.

its publication (1871) in maintaining that the nuclear cells and not the posterior roots were the primary seat of the affection, and that this was inflammatory rather than purely atrophic: I think that if he had lived till now he would have acknowledged the soundness of my criticism. But the merit of establishing the central seat of neuralgia was all his own ; and he thereby rescued the malady from being a mere appendage to gout and rheumatism and debility, and gave it a definite place among the neuroses.

Neuralgia is the manifestation of a discharging lesion at the origin of a sensory nerve : it is, as another has called it, a nerve-storm. Sometimes its lightning may come in single flashes, as in the terrible tic-douloureux — the "epileptiform neuralgia" of Trousseau ; sometimes in a more continuous sheet of flame, as in the ordinary type ; sometimes, as in migraine, there may be a complex variety of lightning, and thunder, and rain, and wind. But, however it may appear, it is a storm — an electric outburst. Its peculiarity stands in its absolute localisation — its confinement to the nucleus of one sensory nerve on one side of the body. Why it should make this selective choice, we cannot say : what most concerns us is that drugs do the same thing. "Aconitine" says Schroff "produces a peculiar feeling of drawing and pressure in the cheeks, the upper jaw, the forehead — in a word, in the parts supplied by the trigeminal nerve. This feeling increases little by little in intensity, and is transformed at first into a remittent pain which shifts its place, later into a continued pain of considerable intensity." Why this result in the trigeminus and not elsewhere? The drug, in Schroff's experiments, was swallowed, absorbed into the blood, carried about in the circulation. Its molecules reached the nuclei of all the sensitive nerves alike : why should that of the fifth only undergo the morbid change which makes it the seat of pain? We know not ; but this we know, that when the same nerve is the seat of neuralgia, of such kind as to indicate *Aconite* and its alkaloid, they are as effective to cure as they are to cause it.

We have not the same evidence for the neuralgia-producing power of *Belladonna* as we have for that of Aconite, but its curative action is no less decided. It is by pathological analogy, rather than by the facial pains experienced by one of Dr. Hale's provers of Atropia, that we must argue its homœopathicity. Neuralgia is a quasi-inflammatory condition of the gray cells at the root of a sensory nerve. Belladonna sets up a similar morbid process in the gray matter of the cerebral hemispheres, and therefore can only act as a similar here. The small dosage required bears out the argument ; and here also the trigeminus is the seat of the neuralgia in the great majority of the cases it benefits.

In Aconite and Belladonna we thus have two great remedies for neuralgia affecting the fifth nerve, and they are types of two classes of analogous remedies. With Aconite anæsthesia accompanies the pain, with Belladonna hyperæsthesia. It is hardly likely that an irritation of sensory cells so similar as in either case to eventuate in neuralgic pain should in one produce diminished, in the other increased, sensibility : it is more likely, therefore, that these conditions are developed farther back and higher up than the pain, i.e. in the sensory centres. However this may be, Arsenicum and Platina follow Aconite in having numbness or actual loss of sensation with their nerve-pain ; while Chamomilla, China, Nux vomica, and of course Hyoscyamus and Stramonium, stand with Belladonna. Hydrocotyle has caused hyperæsthesia of the fifth without pain ; while Colocynth, Iris and Spigelia have developed neuralgic suffering without any other modification of sensibility.

3. Our drug-action will become yet more cerebral in its localisation if from the seat of trigeminal neuralgia we pass to that of *migraine.* So much is owing to the French physicians for their study of this malady that their form of the Greek *hemicrania* may well pass current among us, though its old English shape of "megrim" has much claim to adoption. It is under this last name that Dr. Robert Liveing has described it, in one of the best monographs with which I am acquainted in the whole range of medical literature.[1]

Migraine, in its most common form, is known as "sick-headache," from two of its most constant and distressing features. Some persons, for similar reasons, describe it as "blind-headache." Dr. Liveing, however, studying the malady *more Hahnemanniano,* has been able to construct a complete picture of its phenomena, seen in full only in the most typical cases, but so occurring by one, or two, or three as to

---

[1] On *Megrim, sick-headache, and some allied disorders.* London, 1873.

leave no doubt of their coherence one with another
and with the essential disease. "The forms of megrim" he sums up "range from the simplest hemi-
cranial pain, transient half-vision, or sick-giddiness,
to cases which present a complex assemblage of phenomena and wide range of sensorial disturbance." In
a well-developed example, the attack is ushered in by
a peculiar disturbance of vision, and culminates in
headache and vomiting; but during the culmination
there may occur numbness and tingling on one or
other side of the body, and disorder of speech or
thought, and at any time throughout the attack there
may be vertigo, hyperæsthesia and hallucinations of
the other special senses, and emotional disturbance,
especially "a vague and unaccountable sense of fear."
The face is generally pale and sunken; the heart is
slow and the pulse contracted.

Dr. Liveing, writing in 1873, was led by the physiology of his day to find the primary seat of this
"nerve-storm" in the optic thalamus. That of ours
would bid us look farther; and it is significant to
find him noting that when disorder of speech was
present, and numbness and tingling of one arm co-
existed (as it very often did), it was always the right
one. We, I say, should look to the cerebral hemispheres — in their sensory area — as the seat of the
phenomena: we should say that the "storm" began
in the angular gyrus, and thence spread along the
other sensory centres to that of speech, at last becoming externalised pain with its sympathetic nausea and vomiting.

But, however this may be, our great object is to utilise the picture thus before us, and find medicines truly similar to its essential features. Let us take those of vision as a starting-point. The affection is a blind spot, most frequently central, but sometimes assuming the form of hemiopia; and then almost always lateral, very rarely superior or inferior. The blur is dark against a bright ground like the sky, but luminous on closing the eyes; and is generally surrounded with zigzag coruscations, often compared (in shape) to the bastion-work of a fortress. It spreads peripherally or laterally, according as it is central or hemiopic, and the vision clears at the primary spot as the obscuration widens. Its course is a brief one, and then comes the headache. It seems to be bi-lateral in all cases, and to be unconnected with any change (appreciable by the ophthalmoscope) in the retina.

In searching for remedies with this clue, we are first of all led to *Ignatia*. Hahnemann observed, sixteen hours after taking a dose, "a circle of brilliant white, glittering zigzags beyond the visual point when looking at anything, whereby the letters on which the sight is directed become invisible, but those at the side are more distinct;" and again he notes, after thirty hours, "a zigzag and serpentine white glittering at the side of the visual point, soon after dinner." In a note, he directs attention to these symptoms as "very much resembling Herz's so-called spurious vertigo." I cannot trace the reference; but should think it most probable that Herz was describing the visual phenomena of migraine, of which giddiness is often a potent element. Looking then to the other features of drug and disease, we find that the headaches caused by Ignatia were frequent and severe, though only once associated with inclination to vomit ; that difficulty of thinking and speaking was noted by two of Jörg's provers of it ; and that hyperæsthesia of the special senses and emotional disturbance are very characteristic of it. Ignatia, therefore, would be well indicated for migraine beginning with central blur and coruscations, and going on to severe pain with such phenomena as those just mentioned. It would be the more suitable if the patient were of impressionable temperament ; if the attacks were specially liable to be provoked by emotion ; if the pain assumed the form of "clavus ; " and if the paroxysm passed off with the emission of a quantity of limpid urine.

Nux vomica also has produced the visual phenomena which Hahnemann compares to the "vertigo spuria" of Herz ;[1] so that it would seem as if the Strychnia common to the two were its real exciting cause. In its pathogenesis, however, though heroically enough obtained, this symptom has not appeared ; and our wisdom will be, for the present at least, to use the matrix drugs. We thus moreover get two remedies instead of one ; for you know that Nux vomica and Ignatia have many points of distinction. The patient whom the former suits is one of different temperament and habits from those which call for the

[1] *Mat. Med. Pura,* tr. by Dudgeon, S. 145.

latter; and (remembering how it is indicated for
brain-workers) it is noteworthy how many men of
high intellectual power — Woolaston, Herschel, Airy,
Lebert, Du Bois Reymond — have furnished narratives of their personal experience with migraine to Dr.
Liveing's book. The Nux vomica migraine would,
from its pathogenesis, have more vertigo about it
than that of Ignatia, as much hyperæsthesia, but less
strictly emotional excitement, — if anything of this
kind were disturbed, it would be what we call the
"temper." Errors in diet might well be its exciting
cause; but I do not think that any stress must be
laid on vomiting in the course of it, as only once has
the headache of Nux had this concomitant, and then
it came on after dinner, and was sour, — very different
from the way it occurs in migraine.

When first the visual symptoms of migraine were
definitely described in our day, they recalled to several the results obtained by Purkinje with *Digitalis*.
Experimenting on himself on two occasions with the
extract and infusion, he both times noticed much
flickering before the eyes, and on one occasion thus
describes his experiences: — "The figures which the
flickering formed were called flickering roses, because
the form of the rose was the type: in the first experiment there were round spots in the field of
vision, the space circumscribed by four deep oval
lines forming four large convex indentations, and
the waves of light and shade surrounding this exhibited the same form only less indentated. The flickering figures which appeared on the second day, and

which reached their height on the third; were surrounded by curved lines with five indentations, but more superficial, which were again surrounded by enlarging waves of light and shade with superficial indentations. During the latter days, when the flickering decreased, there were noticed only fragments with the rose formation on the side, like the small segment of the larger and more superficially indentated circle." Now one may agree with Dr. Liveing that the resemblance between these "roses" and the "fortification pattern" of migraine is not striking, but it is near enough to call attention to the drug, and on looking farther we find in its visual and other symptoms a close parallel with the disease. In a patient of Baker's taking it muscæ volitantes were seen before the eyes on looking at distant objects, which, when the eyes were covered, became luminous. Dr. Brunton, when proving Digitaline, saw a large bright spot advancing before him; and our own Baehr, under the same circumstances, had the upper half of his field of vision covered with a dark cloud. In him, moreover, a parietal headache set in in the morning, became worse in the afternoon, and "increased in the evening to a violent migraine." This was not indeed an unprecedented occurrence with him, but it was different from his ordinary attack in that then it was always at its worst in the morning on rising. Headache, moreover, often severe, is a frequent symptom of both Digitalis and its alkaloid: vertigo is no less marked from it; and its vomiting is of cerebral origin, slow of coming on, but, when excited, severe and long lasting. Remember also the slow pulse of migraine, and the pale face and contracted arteries — the last being so often a marked feature as to lead Du Bois Reymond to suppose the disorder a vaso-motor neurosis; and you have in forms of it frequently appearing a complete picture of the effects of Digitalis.

Again, of the Austrian provers of *Cyclamen*, seven had more or less obscuration of sight, and four had flickering before the eyes. One of these, whose eyes were "weak" and required glasses, had this symptom — after two doses of the drug — for six days in the right eye, for three weeks (though less severely) in the left. It began, too, with violent headache, which lasted unchanged for two days, diminished on the third, and disappeared on the fourth. At one time he speaks of seeing a "luminous ball" before the eyes: at another, "with the eyes opened or closed, he seemed to see at a distance of about two feet a dark disk as large as a two-groschen piece, which seemed frequently to be pierced by brilliant lightnings." Vertigo and mental confusion appeared in the provers, and have been verified by a good cure. In this case they occurred in a woman at the climacteric; and Dr. Eidherr had long before given us several cases in which the head and eye symptoms of the drug had co-existed with catamenial derangements such as it causes, and which had yielded to it. Cyclamen therefore should be useful in migraine occurring in such subjects and under such circumstances; and especially where its character was such as to lead to its being called "blind-headache."

The last medicine I have to mention in connexion with this malady is the *Iris versicolor*. I have nothing to add to what I have written of it in the last edition of my *Pharmacodynamics;* but as it was the "blur before the eyes" preceding a sick-headache which first led to its employment in true migraine, it could not be omitted here.

I have made no attempt, in this little study, to mention all the remedies applicable to migraine. I know the value in it of Belladonna, Calcarea, Sepia; of Sanguinaria, Silica, Stannum, and of Zincum sulphuricum. My aim has been simply to show how the finer pathological study of the present day can be utilised in our practice, by associating therewith the knowledge we possess of the physiological action of drugs. If sick-headaches are at all as common with you as they are on our side of the water, you will not regret the time spent on the study of some less common, but not less promising, remedies for the aid of their victims.

4. Finally, the whole subject of the "aura epileptica" must be studied in its relations to drug-action. There is in Dr. Hammond's *Diseases of the Nervous System* a copious list of these, taken from a French source, and arranged with regard to their apparent place of origin. No better subject could be taken for a graduation thesis than a comparison of this list with the Materia Medica, and an exhibition of the results. It might fairly tend to make our therapeutics of this obstinate malady more uniformly and permanently successful.

II. I pass now from forms of disease to individual medicines, noting any relation their action may have to the localised functions of the cerebral cortex.

1. Of *Aconite* I have said enough in relation to trigeminal neuralgia ; but its anæsthesia yet demands a word. Liegois and Hottot, as also Ringer and Murrell, have from their experiments come to the conclusion that it produces this effect by acting on a supra-spinal sensory centre. Hitherto this has been assumed to be the optic thalamus ; but now the tactile cortical centre may lay equal claim to being the seat of its influence. And what lends support thereto is the marked emotional excitement characteristic of the drug. Hahnemann's "anxious impatience, unappeasable restlessness, and agonised tossing about" will be remembered, and also the fear of death well known to indicate it in many disorders. Dr. Samuel Potter relates[1] a partial proving of the drug which will illustrate this. In the course of the well-known Milwaukee experiments, he was supplied with ten undistinguished vials, one of which contained Aconite in the 3rd cent. dilution, and he was to identify it (if he could) by its distinctive effects. He took frequently repeated doses from one of them, in a short time began to experience agitation, anxiety (wholly causeless) about an absent relative, and finally such impressionability that the least thing made him start, work became impossible, and he had to go to bed. A similar trial of the contents of the other vials produced no such effects, and he decided (rightly) on the first as the medicated one.

[1] *Hahn. Monthly*, Sept., 1880.

2. *Belladonna* plays so large a part in nervous disorders because it induces inflammatory irritation of
the gray matter of the nervous centres. This doctrine — which was only a clothing in the language
of modern pathology of a conception universally
current in the school of Hahnemann — has been
supported by the researches of Dr. Harley. He
rejects the terms "narcotic" and "sedative" hitherto in use to describe the neurotic influence of the
drug, and substitutes "excessive stimulation of the
nerve-centres, attended by increased oxidation" as
accounting for the whole of the phenomena produced by it. As the hyperæsthesia it causes in the
sensory sphere and its jactitation in the motor are
accompanied by ideational disorder in its delirium,
there seems no doubt of its action being on the cortical centres. I need not remind you of the number of therapeutic applications which this action
explains.

With Atropia, the so-called "active principle" of
Belladonna, the inflammatory character of the symptoms is much less pronounced, though it is there.
Michéa, who gave it largely to epileptics, compares
the symptoms it produces, in cautiously-increased
doses, in the muscles of articulation and of the extremities, to those which occur in the early stages
of the general paralysis of the insane; and relates
a case in which the diagnosis between these two
interpretations was difficult.

3. *Cannabis Indica* — the hemp of Hindustan —
developes a peculiar resin known as haschisch, which

is eaten or smoked throughout the East (the practice is not, I have lately read, unknown in the West) for its narcotic qualities. The intoxication it induces is characterised chiefly by (as I have written) "an intense exaltation, in which all perceptions and conceptions, all sensations and emotions, are exaggerated to the utmost degree. Distances seem infinite and time endless; pleasure is paradise itself, and any painful thought or feeling plunges at once into the depth of misery. Hallucinations of the senses are common; and the least suggestion will set going a train of vivid mental illusions." That the seat of the drug's action is the cortex cerebri is pretty evident; but is confirmed by there often being an accompanying headache, with sensation as of the brain boiling over, and lifting the cranial arch like the lid of a tea-kettle. Now you may be aware that in the disease just mentioned — "general paralysis of the insane" — one of the earliest mental symptoms is a morbid exaltation of the ideas, leading to notions and schemes of extravagant grandeur. The lesion seems to be an inflammation of the gray matter of the hemispheres and of the neighbouring membranes. It is not certain in which the mischief begins; but even were it in the meninges, the cerebral irritation would be like that which causes contracture in lateral sclerosis, and might be greatly modified by the remedy. I would add that Purkinje experienced a similar exaggeration of the perceptions of time and distance from *Nux moschata*, and that the neurotic effects of this drug — which have as yet hardly been utilised —

correspond well to a somewhat more advanced stage
of the malady.

4. I did not mention *Cocculus* while speaking of
migraine, because it does not cause the visual symptom of the disorder, for which, nevertheless, it is a
leading remedy. The fact is, it covers its motorial
side alone, Cocculus having apparently no action on
the sensory tract of the cranio-spinal axis. It is
therefore indicated when migraine occurs under the
form of "sick-giddiness" only, and when vertigo and
vomiting (often with slow pulse) form the leading
features of the more fully-developed attack. I have
given in my *Pharmacodynamics* a beautiful cure of
Dr. Black's illustrating its action, which would appear to be exerted mainly on the cerebellum and its
associated centres, as shewn by the character of the
convulsions excited by its alkaloid picrotoxine. These
are semi-circular and backward movements, and roll-
ings over on the axis of the body, such as have been
observed from experimental injury to the crura cere-
belli (not "cerebri," as misprinted in my book), and
from electrisation of the cerebellum itself.[1]

5. This, as well as alphabetical order, leads me
here to speak of the influence of *Conium* over the
optical part of the mechanism of equilibration. That
the visual impressions proper for this function may
be conveyed to the cerebellum, it is necessary that
the axis of vision as preserved by the ocular muscles
should be correct. Conium paralyses these, and
hence the giddiness on every fresh adjustment of

---

[1] Ferrier, *op. cit.*, p. 98.

focus observed by Dr. Harley, and occasionally such symptoms as those experienced by Dr. Edward Curtis. After taking half a drachm of Squibb's fluid extract, he could not walk across the room *with his eyes open* without giddiness, reeling, and feeling as if he would vomit ; but directly he closed his eyes all the symptoms passed off, and he could move safely.[1] This explains the benefit obtained in threatened sea-sickness from shutting the eyes, or at any rate not looking at the swaying boat and waves. In this connexion let me remind you that *Quinine* and *Salicylic acid* exercise a similar influence as regards the part played in equilibration by the semi-circular canals of the ear ; and that here the pathogenetic effect has been turned to good account, both remedies having been used with advantage in Menière's disease in its curable stage.

6. The emotional erethism caused by *Chamomilla* and *Coffea* (with some other vegetable medicines) on the one side, and by *Iodium* and *Mercurius* on the other, seems referable to action on the sensory region of the cortex, for in all it coexists with great sensitiveness to pain. With the former it is functional only, but with the latter (especially with Mercury) it may be the beginning of organic disease, i.e. of softening. It is worthy of note that insomnia characterises all. Sleep, whatever be its cause, is certainly a function of the brain proper: only consciousness can become unconscious, as only life can die. The mental condition induced by Iodine and Mercury is

[1] Allen, *sub voce.*

little known, and should be studied: it may often guide us to these profoundly acting drugs at the commencement of serious brain disease, and arrest it ere it has gone into substantive change.

7. And now a few words on *Santonine.* Everyone knows the xanthopsia — the yellow vision produced by this drug, and you are probably aware that in higher degrees of its influence the colouring is violet. These, however, are visual *illusions*, which disappear with the objects on closing the eyes, and are due to alterations in the pressure of blood at some part of the optic tracts. But Santonine produces also visual *hallucinations*, seen just as well with the eyes shut, and independent alike in time, in frequency of occurrence, and in degree, of the illusions. Dr. Edmund Rose, of Berlin, who has studied these phenomena very fully, shews that they must be of cerebral origin; and points in confirmation to concomitant hallucinations of other senses — touch, sight, taste, *not* hearing, with the evidently sympathetic vomiting, and the commencement of the convulsions (when they occur in poisoning by it) in the muscles supplied by the cranial nerves. A fellow-professor has spoken of experiencing some mental incoherence from it; and Dr. Farquharson, of London, who tried it on himself, writes — "The best-marked symptom was a feeling of profound and most unusual depression, accompanied by so much irresolution and want of confidence in my own powers as to render me quite unfit for work of any kind. This invariably followed even a single five-grain dose,

and, beginning with dulness and heaviness, ran on into very much the sort of melancholia which I imagine jaundice must produce."

This marked action on the brain ought to be utilised.[1] That it can be appears from some of the eye cases treated by Drs. Ogston and Dyce Brown, among which cerebral amblyopia and paralysis of the motor oculi were greatly benefited, and concomitant headaches removed. It evidently acts on the sensory portion of the cortex cerebri.

As this will be the last time we shall study individual drugs together, I will ask your attention to one feature in our mode of proceeding which I think worthy of notice. It has been very common of late to hear it said that drugs "have" such and such symptoms, or that we find these "under them;" without any heed being taken as to how the drugs got the symptoms, or how the symptoms came under the drugs. Now this practice may be convenient, for brevity's sake;—but I venture to object to it as perilous. Had we a pure and sound Materia Medica, it might be indulged in; but in our present confusion of genuine and imaginary, of pathogenetic and clinical, of real experience and wholly supposititious inferences, it is most unsafe. Our only security lies in ascertaining the source of any asserted drug-action

[1] Dr. Heber Smith, who was present at this lecture, afterwards informed me that he had obtained striking and permanent benefit from Santonine in a case of recurring occipital headaches of much severity, accompanied by visual hallucinations as of red fire-balls moving about. He gave the 2 x trituration.

we wish to utilise, and — if we are writing for or teaching others — in stating it. To do so may often involve periphrasis ; but this is one of the cases in which the shortest cut is the longest way. To make sure of our ground before we stand on it is the wisest course to follow when we are threading a morass.

And so we are brought back to the state of our Materia Medica, and the need of reform therein, which was broached in our fourth lecture. In the next and concluding one I purpose taking up the sub·ject anew, and shewing more in detail what is our present position, and what our prospects and means of remedy.

XII.

THE FUTURE OF PHARMACODYNAMICS.

As an introduction to the third volume of his *Materia Medica Pura* (1st ed. 1816, 2nd ed. 1825), Hahnemann prefixed "An examination of the sources of the common Materia Medica." In this scathing criticism he passes in review the current methods of arriving at the knowledge of the value of medicines, and exposes their utter inefficiency. He discusses the ascription of general therapeutic virtues, as when drugs are styled resolvent, tonic, and so forth; the inference from sensible properties, as those of the bitters and aromatics, or from chemical qualities; and the *usus in morbis*, shewing conclusions from this source to be vitiated by polypharmacy and lack of individualisation. Such sources, he concluded, were turbid, and a Materia Medica drawn from them was unworthy of confidence. It was full of error and falsehood; and "error in the most serious and important of all earthly vocations, the healing of the sick, must have the most grievous consequences;" while "falsehood here is the greatest of crimes, being nothing less than high treason against humanity."

259

These were strong words, but they only expressed what Hahnemann sincerely felt. Nor had he confined himself to feeling, or to shewing the worthlessness of the existing material. He had learned and taught the more excellent way of arriving at the knowledge of medicines, viz. : by ascertaining their action on the healthy body ; and he had made some progress in such ascertainment, and in communicating its results to the profession. Before he died, he had published ten volumes containing pathogeneses of drugs — six of the *Materia Medica Pura*, four of the *Chronic Diseases*. The impetus he gave to the work of proving drugs has continued to this day among his own disciples, and has spread even among the followers of traditional medicine, so that his ten small octavos, with their ninety three medicines, have expanded into the ten large ones of Allen, with their seven hundred. Our quantity of material is unquestionable, — but it is necessary that we keep strict watch over its quality : ever and anon, if we would follow Hahnemann, we must institute an " examination into the sources of the common Materia Medica."

The master's way of proceeding, in his day, was to plunge at once into criticism. In ours, we shall do more wisely to begin by laying down principles, which shall serve us hereafter as standards for judgment. Assuming, then, that the best way of learning the action of medicines is the investigation of their effects in health, we have two classes of organisms on which these can be ascertained — the lower animals, and man himself.

I. There was a time when the *corpus vile* of brutes was thought the only ground on which *fiet experimentum*; and even now it plays by far the largest part in the pharmacological research of the profession at large. If this were sound practice, Hahnemann would be somewhat discredited; for he, recognising that it was available, deliberately rejected it. But have his arguments against its adequacy ever been answered? The first is that the effects of drugs are different on them and on us, and different as between themselves. "A pig can swallow a large quantity of nux vomica without injury, and yet men have been killed with fifteen grains A dog bore an ounce of the fresh leaves, flowers and seeds of monkshood: what man would not have died of such a dose? Horses eat it, when dried, without injury. Yew leaves, though so fatal to man, fatten some of our domestic animals. . . . The stomach of a wolf poisoned by monkshood was found inflamed, but not that of a large and a small cat, poisoned by the same substance." Thus Hahnemann, and similar facts have come to light in later times, among which I may mention the impunity with which the rabbit may be fed for days upon belladonna leaves. The argument from them has been urged afresh in the forty-first volume of the *British Journal of Homœopathy*, and shewn to be borne out by the contradictory results of recent pharmacological research on animals.

The second objection is yet more destructive: it is that we cannot obtain subjective symptoms from dumb creatures, and we have seen how important

these are in the knowledge — for curative purposes
— of disease, and therefore also of drugs.   We may
point this objection by the instance of Aconite.   In
experiments on animals, loss of sensibility of the
surface is often noted: hence the drug is supposed
to be an anæsthetic, and suited for employment in
neuralgia and other simple pains, for which it must be
given in physiological doses, or — when the affected
parts can be reached — applied locally.   But con-
sult human poisonings, or, still better, provings, and
another tale is told.   While the surface may be insensible to external impressions, it is not so to the
patient's own consciousness   It is an *anæsthesia do-
lorosa* from which he is suffering, and one which — as
we have seen in Schroff's provers — may develope
into actual neuralgic pain, to which therefore Aconite
is truly homœopathic, and which it will cure by internal administration and in non perturbing dosage.

These objections are surely fatal to any exclusive
or even predominant reliance on experiments upon
animals, for ascertaining the properties of drugs.
But on the other hand they have a place, which
Hahnemann was ready to acknowledge (thirty years
before Majendie began their systematic institution),
and which the provings of his school, when thorough,
have always given them.   Besides the induction of
the more violent effects of the drugs, and the ascertainment at will — by autopsies — of the lesions they
set up, we can learn upon these subjects the result of
their long continued employment in doses sufficient
to change without killing.   In this way Wegner has

discovered the power of Phosphorus to induce a plastic irritation of periosteum and of the interstitial tissue of the stomach and liver, and Eugene Curie has shewn Bryonia capable of exciting pseudo-membranous deposit and Drosera that of tubercle. Again, experiments on animals lend themselves to analysis and interpretation. Dr. Lauder Brunton has well shewn how in this way the rapid circulation of Atropia has been proved to be due to paresis of the terminal extremities of the vagi in the heart; and the opposite effect of Digitalis has been demonstrated to result from stimulation of the same inhibitory fibres at their origin. It is not always that here *le jeu vaut la chandelle* — that we take much by our knowledge; but, assuming it to be worth having, it is certainly from experiments on animals that we must obtain it.

II. Such experiments, then, being of subsidiary value only, we turn to the action of drugs on the human body as the main source of our knowledge of them. This knowledge must be gained in pharmacology, as in all other sciences, by observation and by experiment.

1. Observation, in the present instance, has for its field poisoning of healthy and over-dosing of sick persons; and each of these sub-divisions requires separate discussion.

a. Poisoning is obviously limited to the comparatively small class of drugs sufficiently virulent to produce such effects. Here, however, it is of great value. It supplies the more violent disturbances and

the *post mortem* changes induced by medicinal substances better (because more surely) than experiments on animals can do : it aids us greatly in arriving at the lesions they can produce, and in obtaining
correspondence of seat between drug-action and disease. Records of poisonings and works on toxicology have therefore been always largely employed,
from Hahnemann downwards, in the construction of
our pathogeneses; and nothing can be said against
this source of knowledge save that it is, as it were,
illegitimate. Poisonings are the product of crime or
of carelessness, and in the progress of society should
become more and more rare; so that we may not
lean too confidently upon them as materials of future
information.

b. Over-dosing may also be said to be a remediable
error ; but as long as traditional medicine is practised it will be liable to occur again and again, as it
has occurred in the past. The object of both antipathic and allœopathic medication being to induce the
physiological actions of drugs, these are continually
being observed; while even in "alterative" treatment
the ponderable doses deemed necessary, and the occasionally quick susceptibilities of patients, determine
the eliciting of collateral effects. The older treatises
on Materia Medica draw largely on such observations,
partly for knowledge as to pathogenetic action, and
partly for warning against excessive dosage. The
obvious weakness of effects so obtained is the uncer-
tainty which belongs to them, owing to their exhib-
iter being already the subject of disease. Of course,

if this be of a definite and limited character, and consisting with fair general health ; and if all symptoms
conceivably resulting from it, or occupying the same
seat, are excluded, and likewise all phenomena previously observed in or by the patient during his illness, — then pathogenetic effects may be taken from
the sick almost as safely as from healthy persons.
Some of our best records of the symptoms produced
by atropia, as those of Grandi, Michéa, and Lussana,
have been in this manner obtained ; and without it
we should know next to nothing of the physiological
action of the bromide and iodide of potassium and of
salicylic acid.

2. We come now to experiment, which here, as in
other departments of research, should be our principal resource. Very little use, however, had been
made of it up to the time of Hahnemann. Haller's
insight had perceived its need ; and he had written
— "It is upon the healthy body first that the medicine, free from any foreign admixture, is to be tested ;
its taste and odour to be ascertained, and then, small
doses being swallowed, their effects to be fully noted,
how the pulse behaves, how the temperature, how
the breathing, how the excretions." But his words
had fallen on barren ground ; for there is no trace of
any connexion between them and the few provings
which were extant at the end of the last century.
Stoerck (1750–1760) had swallowed a few doses of aconite, conium, and colchicum, — merely, however, to
ascertain whether and how far they could be administered with impunity. Alexander (1768) had tested

on his own person castor, saffron, nitre and camphor; but here again as much to try whether these substances had any activity at all (which question in the case of the first two he was led to answer in the negative) as to discover their "doses and effects" if really operative. Grimm (1767), Crumpe (1793) and Bard (1795) had made some experiments with opium, Coste and Willemet (1778) some with asarum, and Wasserberg one with belladonna. These were the only forerunners of Hahnemann ; and how few and feeble were their efforts! He, on the other hand, once persuaded of the necessity for therapeutics of drug-provings on the healthy human body, proceeded to institute them on the most extensive scale, and to publish his results.

Before we go on to estimate the worth of these, let us consider how a proving should be conducted. It should secure healthy persons as its subjects, or such as, from the limited seat and known symptoms of any malady they might have, are practically so. They should either have some medical knowledge themselves, or should be interrogated by a superintending physician. They should be free from the action of any disturbing causes, mental or bodily. Since acute disease is generally induced by single potent causes — a chill, a shock, a morbid poison, &c., and chronic disease by a succession of slighter ones, the medicine to be tested should be taken both in single full doses, and at another time in small ones repeated more or less frequently until some effect is obtained. To avoid the influence of expectant attention, the prover should not know what drug he is taking. Finally, that mere natural oscillations in health should not be set down to the medicines, the prover should note any symptoms observed in himself for some days previous to beginning his experiments. This last refinement must not be expected of Hahnemann, however much its absence may vitiate his results. It was not till 1871 that Dr Hamilton shewed us[1] how many slight deviations from the norm will occur in a man presumably healthy who records his own daily feelings and doings ; nor till 1877 that your own Professor C Wesselhoeft exhibited the effect of combining herewith the influence of expectant attention.[2] Putting this aside for the moment, let us enquire how far the provings published by Hahnemann come up to the standard we have erected. Parts of what I now have to say have already seen the light in divers places ; but I think it well to bring the facts and considerations relating to the subject in one focus to-day, that, knowing what pharmacodynamics has for you in the present, you may follow me in the attempt to state its needs for the future.

1. Hahnemann's first publication (1805) was his *Fragmenta de viribus medicamentorum positivis.* It was written, as its name implies, in Latin, and contained pathogeneses of twenty-seven drugs. As its contents were subsequently incorporated with the larger works — mainly with the *Materia Medica Pura,* I will only say of them that the symptoms referable

---

[1] *Brit. Journ of Hom.,* xxix., 565.
[2] *Trans. Amer. Inst. of Hom.,* 1877.

to provings appear to have been the results of single full doses of the several drugs, and that Hahnemann speaks of them as having been noted "in a careful and sceptical manner."

2. The *Reine Arzneimittellehre*, better known by its Latin name of *Materia Medica Pura*, began to be issued, in successive volumes, in 1811, and the first edition was completed in 1821. In 1822 a "second, augmented edition" began to appear, and ran on in like manner until 1827. In 1830 and 1833 respectively the first and second volumes entered a third edition, which, however, went no farther. Our translation of the work thus contains the medicines of the first two volumes in their third, of the other four in their second revision. The dates of these are important to bear in mind, as they correspond to epochs of Hahnemann's life, and afford a clue to the manner in which the symptom lists were obtained.

The first edition of the *Materia Medica Pura* contained sixty-one medicines, besides the magnet (of which the less said the better, though the numerous citations from authors ' shew that Hahnemann was not alone in supposing it to be capable of disordering the healthy functions). In the second edition, three more — Ambra, Carbo animalis and vegetabilis — were added to the sixth volume (1827) ; and, in the third, Causticum was omitted from the second volume. The total is thus sixty three. The pathogeneses are made up of three elements, each of which must be examined separately.

---

¹ In the second edition, 195 (out of 861).

a. The first are the symptoms vouched for by
Hahnemann himself, and which, until the third edition, occupied a separate list, — the symptoms of fellow-provers and the extracts from medical literature
appearing together in a second as "Observations of
Others." Those of the *Fragmenta de viribus* were
elicited, Hahnemann tells us, by experiments made on
his own person and on that of others whom he knew
to be perfectly healthy — probably members of his
family. We may reasonably so account of the
symptoms furnished by him to the *first* edition of
the *Materia Medica Pura ;* and, as they had probably
been obtained during the six years intervening between the two publications, we may suppose these
likewise to be the effects of single full doses.

The same cannot be said, however, of the additions made by him to the second and third editions.
Even for those of the earlier volumes, it is hardly
likely that he had instituted fresh experiments with
the same drugs on himself or his family ; and those
made on his disciples — including his son Friedrich
— are separately given. There is reason to think,
therefore, that he had already begun his undoubted
later practice of noting fresh symptoms appearing
in patients after drugs had been given them, and
ascribing these to the medicines they were taking.
During this period, i.e. from 1811 onwards, he was
using attenuated remedies pretty freely, so that
many of the symptoms thus obtained must have
appeared while these were being administered.

We shall see immediately what grave objections

lie against this mode of eliciting the supposed pathogenetic effects of drugs.   Its large use by Hahnemann vitiates the symptoms supplied by him even
to the *Materia Medica Pura*.   It would be a boon
if any one would have the patience to go through
our present English version with the first edition of
the original, and mark all symptoms belonging to it,
which then we might count on as genuine.   Until
any have been so identified, we cannot be sure of
their not belonging to a different category.

b. In the volumes from the second (1816) onwards,
Hahnemann was able to follow up his own symptoms
with a number obtained by fellow-provers, of whom
thirty-seven in all co-operated with him during the
progress of the work.   They were disciples of his
who had gathered round him during his brilliant
career at Leipsic, and whom he had enlisted in the
task of proving.   From his own account he appears
to have taken every precaution that they, and any
others on whom they might experiment, should be
in good health and under normal circumstances ; and
that the influence of any disturbing causes should
be eliminated by omitting, or at least bracketing as
dubious, symptoms then occurring.   Dr. Hering
also tells us how carefully he examined his provers
on receiving their day-books, so as to ensure fulness
and accuracy.   As regards the manner of proceeding, it would seem that insoluble substances were
proved in the first trituration (centesimal of course),
and vegetable drugs in the mother-tincture — repeated small doses being taken until some effect was

produced. An exception must be made for the three new medicines of the second edition of the sixth volume — Ambra, Carbo animalis and vegetabilis, — where, certainly in the case of the last, and probably in that of the first two, the third trituration was employed.

I think, then, that — all deductions being made for physiological oscillations and the influence of expectant attention — the symptoms of Hahnemann's fellow-provers in the *Materia Medica Pura* may be accepted as substantially genuine and trustworthy.

c. We come now to symptoms taken from authors, which are absent from only thirteen of the pathogeneses, and with which the remainder are so abundantly supplied as to raise the total to over four thousand, about one-twelfth of the whole. These symptoms are either general statements of writers on Materia Medica, or observations of poisoning or over-dosing. The first may stand for what they are worth ; but, considering the uncritical character of most of their enunciators, they need confirmation before having much weight leant on them. Observations of poisoning are always valuable, provided the *causa* is *vera*, the subject healthy, and the influence of the antidotal treatment allowed for. I make the first two stipulations, because Hahnemann has included in his pathogenesis of Cannabis sativa a number of symptoms observed in workers in hemp, including such trifles as cataract, enlarged liver, ascites, cystoplegia, incurvation of the spine, pneumonia, and mania. Neither by the reporters, nor (I imagine) by any

other person in the world, were or would these perfectly natural disorders be ascribed to the very mild
noxa of the hemp which surrounded their subjects.

I am sorry to say that this is too faithful a specimen of the manner in which Hahnemann has gone
to work with medical literature. He takes a case
or series of cases treated by a drug, and sets down
as many as he pleases of the symptoms noted as
occurring in the patient from day to day as effects
of the same, without regard to the disease which is
being treated or other causes which may have been
operative. In my *Sources of the Homœopathic Materia Medica*, the substance of which is reproduced
in the second and third lectures of the fourth edition
of my *Pharmacodynamics*, I have exhibited many
instances of this mode of proceeding, and have
shewn its disastrous results. I cannot reproduce
them here : I can only give the gross results of an
investigation in which I have consulted every author
and examined every symptom quoted by Hahnemann, both in the *Materia Medica Pura* and in the
Chronic Diseases, so far as those have been accessible
to me and these traceable. I regret to have to say
that many hundreds of these citations, consisting of
symptoms observed in patients, have no more right to
be ascribed to the drugs they were taking than had
the diseases I have just enumerated to be set down
to the hemp in which their subjects were working.

I have shewn (in the same place) by extracts from
his writings that Hahnemann was not insensible to
the risk he was running in drawing from such im-

pure sources; but that the exaggerated notions he
entertained of the potency of drugs led him to set
down all the sufferings, accidents, and changes of
health occurring in patients as solely derived from
the medicines they were taking. He himself, indeed, describes such symptoms as "corroborative"
only; but they have taken their place in his pathogeneses side by side with the others, and thence have
been transferred to manuals and repertories with
their distinguishing references omitted. Nor is it of
much avail to publish corrections, as has been done
in Allen's *Encyclopædia* and in Dudgeon's translation of the *Materia Medica Pura*. "Rage during
the menses" is still quoted as an effect of Aconite,
though it has been shewn that the subject was a
maniac; and so of a host of equally untrustworthy
symptoms.

3. Hahnemann's third and last collection of pathogeneses constitutes (with the introductory essay) the
work entitled *Die chronischen Krankheiten*, "Chronic
Diseases." The first edition consisted of three volumes published in 1828, and a fourth in 1830. The
second and third of these contained fifteen medicines,
all new; the fourth, with two new ones, presented
enlarged symptom-lists of five which had already appeared in the *Materia Medica Pura*. The second
edition appeared in successive parts between 1835 and
1839. Besides the twenty-two medicines of the first
edition it contains twenty-five others. Twelve of
these had already appeared in the *Materia Medica
Pura*, and the rest (with few exceptions) in the *Frag-*

menta or in Stapf's journal, the *Archiv;* but nearly all
have large additions. For all Hahnemann acknowledges contributions from fellow-observers, and for
many cites symptoms from the extant literature of
his day.

Of these last I have nothing to say different from
what has come before us relative to them when occurring in the former work. I have only to speak,
then, of Hahnemann's own contributions, and of those
of his associates.

a. The pathogeneses contained in the two volumes
issued in 1828 appear without a word of explanation
as to how the symptoms were obtained, and the
names of no fellow-observers are mentioned. The
absence of any co-operation on the part of others is
further to be inferred from what we are told of the
first announcement of the work. After six years of
solitude at Cœthen (to which he had been driven from
Leipsic in 1821) Hahnemann "summoned thither his
two oldest and most esteemed disciples, Drs. Stapf
and Gross, and communicated to them his theory of
the origin of chronic diseases, *and his discovery of a
completely new series of medicaments for their cure.*"
So writes Dr. Dudgeon. That he should now first
reveal these new remedies, and in the following year
should publish copious lists of their pathogenetic
effects, confirms the inference to be drawn from his
position and from his silence as to fellow-observers.
He was himself between seventy and eighty years
old, and it is hardly likely that he did anything in
the way of proving upon his own person, as indeed

he makes no mention of having done. We are com·
pelled to the conclusion that he drew these symptoms
from the sufferers from chronic disease who flocked
to·his retreat to avail themselves of his treatment.

The prefatory notices to the several medicines
still further substantiate this view, and throw some
light on the doses with which the symptoms were obtained. He recommends almost all the medicines to
be given in the dilutions from the 18th to the 30th ;
and repeatedly makes some such remark as this —
"For some time past I have given the 6th, 9th, and
12th potencies, but found their effects too violent."
I infer, therefore, that it is these "violent effects" of
the dilutions from the sixth to the twelfth, experienced by the sufferers from chronic disease who took
them, which make up the unexplained pathogeneses
of 1828. Of those of 1830-1839, Hahnemann's con·
tributions — often large — must be explained in the
same manner, save that the 30th dilution must have
been their most frequent (supposed) eliciter, as in
1829 Hahnemann had laid this down as the one form
in which all remedies should be given. I must except,
however, from this description such symptoms as belong to previous provings in which he had taken part,
as those of Cuprum and Mezereum from the *Fragmenta*, and those of Anacardium and Platina from the
Archiv.

It thus appears that (with few exceptions) Hahnemann's contributions to the pathogeneses of the
Chronic Diseases are (supposed) effects of attenuations from the sixth to the thirtieth, observed in sick

persons taking them as medicines. As to their ori·
gin from infinitesimal quantities I have nothing to
say at present. The subject is a large and difficult
one, and demands a lecture to itself, which I have
given to it in my book. This only must be affirmed,
that there can here be no idea of over-dosing in the
ordinary sense of the word. The real question is —
can we trust Hahnemann's discrimination in the mat·
ter of symptoms taken from the sick? Our studies
among his citations from authors supply the answer.
We saw him there, as it were, at work among pa·
tients ; and found that his eager desire for symptoms,
and his over-estimation of the activity of drugs, had
led him in numerous instances to put down as path·
ogenetic effects phenomena which were obviously
those of the disease or of occasional causes. We can
have no confidence, but rather the reverse, that he
has not followed the same course in his observations
on his own patients   Hence (in the first edition of
the *Chronic Diseases*) the thousand symptoms of Calcarea and Phosphorus and the twelve hundred of
Sepia — all derived from sick persons during (prob·
ably) the six or seven years of the Cœthen period.
The re·proving of the last medicine by the American
Institute, in which thirty healthy persons took part,
only yielded 517 symptoms as its result. Hence,
too, the increase of the symptom·list of Sulphur, from
1,041 in 1830 to 1,969 in 1839. The additions are
nearly all Hahnemann's, who certainly was not a
prover at this time, but was (on the ground of the
psora·theory) increasingly a sulphur·giver to the sick.

b. Hahnemann's associates in supplying the pathogeneses of the second edition of the *Chronic Diseases*
were of various kinds. Some of them were published
books, as the *Arzneimittellehre* of Hartlaub and Trinks
and the *Materialien* of Jörg. Some were the men
whose work with him in earlier and better times he
reproduced in these pages. But over and above
these, he acknowledges as fellow-provers for the present work a band of later disciples; and these, it must
be believed, made all their experiments with globules
of the 30th dilution. In the edition of the *Organon*
published in 1833 Hahnemann recommends all provings to be made therewith, as yielding the best results ; and from the preface to Natrum muriaticum
in the *Chronic Diseases* volume of 1830 we find that,
in the case of the three persons who contributed
symptoms to it, the practice had already been begun.
We may fairly extend the inference to all provings
subsequently made.

Here, too, I have no desire to prejudge the question of the pathogenetic activity of infinitesimals,
though certainly any reasonable faith in it is put to
a severe strain. I only wish that you should know
with what you have to deal. I could further wish
that the development of their symptoms herewith was
the worst thing I had to say of Hahnemann's new
associates. But I fear that the demoralising influence of his practice among the sick had begun to
lead them also in the same direction. One offender
here there certainly was — one in many other respects
entitled to sincere honour, Constantine Hering. His

first and (I think) only appearance as one of Hahnemann's associates is in the pathogenesis of Arsenicum,
given as a sort of appendix to the fifth part of the
second edition of the *Chronic Diseases;* and all his
symptoms were avowedly obtained from leprous patients taking the drug, of course in the 30th dilution.
Read them over, as translated from the original in
the *Archiv* in the nineteenth volume of the *British
Journal of Homœopathy* (p. 633), and I am sure you
will agree that they are to be rejected. Wahle is
another of the present band of disciples who seems,
from internal evidence, to have sinned in this way,
and his symptoms must be received with a good deal
of suspicion.

The view of the pathogeneses of the *Chronic Diseases* now put before you was published some ten
years ago, and has re-appeared in more than one form
since. Until quite lately it has received no challenge. Two champions have now advanced to do
battle on behalf of the work, — Dr. Sircar, of India,
and Dr Pope, of England. The former, in a recent
number of his excellent *Calcutta Journal of Medicine,* objects that Hahnemann expressly states that
he learnt the "anti-psoric" character of certain
substances by experimentation on the healthy. This
is so; but the statement in question is taken from
the introduction to the second edition, which includes — as I have shewn — many medicines of the
past and the results of many fresh provings. My
contention that the symptoms were taken from the
sick applies only to Hahnemann's personal and fresh

contributions to the pathogeneses, which nevertheless make up their great bulk. Dr. Pope,[1] while
allowing his inability to traverse my allegations, refuses my conclusions on two grounds, — the viola-
tion of his own canons of which Hahnemann would
have been guilty had he acted as I describe, and the
verification of the symptoms of the work by clinical
experience. To the first I reply, that inconsistency
is too common in human nature for its improbability
to outweigh the evidence I have brought forward.
To the second, that (1) I have already admitted that
the second edition, which is that in our hands, contains many symptoms perfectly trustworthy, and (2)
that too much stress must not be laid on clinical verification, here and there, of such extensive symptom-
lists. Since those of each polychrest contain nearly
every possible variation from health, any success obtained with these drugs would find reflection at some
point of their course. Again, had the curative application of these remedies been drawn solely from
their pathogeneses, it ought to be somewhat commen-
surate with the range of these; whereas Natrum muriaticum, with its 1,345 symptoms, has become an
important remedy, while Natrum carbonicum, with its
1,080, is almost unused, and many similar anomalies
might be presented. The fact is, our use of these
medicines is largely traditional and empirical (the
application, for instance, of Natrum muriaticum to intermittents), and has little relation to their catalogues
of symptoms.

---

[1] *Monthly Hom. Review*, May, 1884.

I have spent much time over these pathogeneses of Hahnemann's, partly from the large space and pre-rogative position they occupy in our Materia Medica, and partly from the typical picture they present of the rest of its material.   Much of this is thoroughly good, — as good as the provings and poisonings of the *Fragmenta* and the first edition of the *Materia Medica Pura ;* and, in its original form, is far better presented.   In this description I include the prov-ings of Jörg, Martin, Schneller, Schroff, Harley and Ringer in the old school, and of the Austrian So-ciety, of many later Germans, of a few French and English, and of perhaps a majority of the numerous American workers, in our own.   On the other hand, Hahnemann's later doings have given the impetus to much work which must be pronounced as thoroughly bad.   Following his example, many symptom-lists have been published as belonging to medicines, with-out the least explanation of the manner in which they were obtained : the *Archiv*, and Hartlaub and Trinks' *Annalen* and *Arzneimittellehre*, are full of them ; and Petroz in France and Hering here gave a number more.   The observation of symptoms in the sick became quite common ; and these patients afforded yet another opportunity for supplying the Materia Medica with symptoms.   Aggravations of their existing troubles occurred from time to time during treatment, and were of course ascribed to the energy of the medicines they were taking ; and these were, not without countenance from Hahne-mann, set down among the symptoms of the drug.

Roth quotes one such instance from the *Chronic Diseases* under Colocynth, and says that he has frequently found the same proceeding resorted to elsewhere. But the imitators of the master's weaknesses have quite outdone him as regards the utilisation of the sick for enriching the Symptomen Codex. When in a prover some existing deviation from health disappeared during the action of a drug, Hahnemann recorded it, adding " Heilwirkung " (curative effect). Only in the instance of Iodium has he done this with definite maladies (as goitre and enlarged glands) treated with the medicine. But his disciples have seized upon the proceeding and carried it to lengths from which he would have shrunk aghast. They have freely admitted " clinical symptoms " (" that abominable fallacy" as Dr. Allen justly terms it " which has poisoned the fountains of our Materia Medica from Hahnemann to the present time ") into our pathogenetic lists, cutting up the cases which have recovered under the action of a remedy into their component parts, and sowing these in the appropriate plots of the schema. They at first (as in Jahr's " Manual ") denoted such symptoms by a sign (° or *) ; but soon grew careless about affixing it, and at last (as in Lippe's " Text-book " and Hering's " Condensed Materia Medica " and " Guiding Symptoms ") avowedly omitted it altogether.

When, side by side with these proceedings, there went on an increase of attenuation up to unimaginable limits, and a corresponding *gobemoucherie* of belief in the energy of these wonderful potencies,

the evil grew ever greater.   It culminated in two con-
summate flowers.   One was Wolf's "proving" (!) of
Thuja, where he took a single globule of Jenichen's
1,000th potency (i.e. about Hahnemann's 40th),[1] and
recorded every deviation which occurred in his health
for the next two years (including an attack of small-
pox) as the result of this violent dose.   The other is
Houat's *Nouvelles Données*, where almost every ill
to which flesh is heir is ascribed (among others) to
those potent poisons, Robinia pseudo-acacia, Cubeba
officinalis, and Piper nigrum.[2]   To such deliberate
vitiations as I have now depicted are to be added
those incidental to time and use — the havoc wrought
by imperfect translation and re-translation, the errors
of repeated copying, and such like.   The result is
(I do not hesitate to say it) that our Materia Medica
is an Augean stable almost as foul as was the common one when Hahnemann exposed its condition,
and set himself to the Herculean task of its puri-
fication.   When we add thereto the unfortunate
schematic arrangement on which I have already
commented, its evil state becomes yet more apparent.   It is, in great part, what some one caustically
styled Jahr's *Manual*, "nonsense made difficult."

What, then, is to be done?   First of all, we must
winnow the many grains of wheat which are scattered through the mass of chaff which lies before
us.   In *pietas* to Hahnemann, his work should (as I
have said) be translated as he has left it, with such

---

[1] See *Brit. Journ. of Hom.*, Jan. 1881, art. " Dilutions."
[2] *Ibid.*, vol. xxvii., 137, and xxxviii., 4.

editing as any similar books would receive. England has done this for the *Materia Medica Pura*, and
she looks to America to do it for the *Chronic Diseases*. When we have these volumes as they should
be presented, let them stand on their own merits.
Let us leave the symptoms of Hahnemann and his
associates, so far as they are only accessible there;
and proceed to deal with the more manageable material which lies outside it. Let this be thoroughly
sifted, — without scepticism, indeed, but without
credulity; and let what remains be presented in the
most accurate, concise, and instructive form, — repetitions being avoided, redundancies pruned, and all
provings given, where possible, in consecutive order,
as related by the experimenters. The student will
then have the text of his medical bible, and will have
it in genuine and intelligible form. Genuine, —
because all versions and copies will be traced back
to their *ultimate* original, and verified, corrected, or
reproduced therefrom; because all clinical symptoms
and (supposed) medicinal aggravations will be excluded, and phenomena observed in patients taking
drugs accepted only on amply sufficient evidence;
because provings themselves will be rigidly scruti-
nised, and not admitted — at least in any completeness or full-sized type — unless their source and
method seem free from objection. And intelligible,
because all observations and experiments will be
related in detail, or in sufficient summary, so as to
preserve the order of evolution of the drug-effects.
We shall then have a series of individual pictures of

the morbid conditions induced by our medicines, and
— so far as they go — shall only have to fit them to
idiopathic disease, on the immortal principle *similia
similibus*, to have the homœopathic method at our
full disposal.

But then will open before us another task, that
of re-proving and fresh experiment. The very exhibition of our extant material will shew many a gap
to be filled up, many a deficiency to be made good.
And further, we ought to improve upon what we
have. With our better knowledge of the natural
history of health, and of the influence of expectant
attention, to obviate self-deception; with our increased command of modes and instruments of precision; with our enhanced acquaintance with the
order we are to disturb and the disorder we desire
to simulate, we ought to make provings greatly
superior to those even of Watzke and his fellows.
For the means of making them I would only say, *Cir-
cumspice.* Dr. Dake[1] calls for Provers' Colleges;
but in the Homœopathic Schools of America you
already have a dozen of such. Let the Materia
Medica classes of these, under their respective Professors, prove each year a single drug; and we shall
soon have a body of experimentation of which we
may be proud.

This, however, is the work of to-morrow. That of

---

[1] The work which this physician has done in exposing the deficiencies of
our existing Materia Medica, and shewing the way to improvement, will always
entitle his name to honour (see Transactions of Amer. Inst. of Homœopathy
for 1857, 1873, 1874, and of World's Hom. Convention of 1876).

to-day is the garnering up of our existing wealth, and the making it immediately available. To this task, feeling its urgency, but knowing its magnitude, I have purposed (God willing) to devote the remainder of my literary life. I have obtained for it the assent and support of the British Homœopathic Society; and I have come over here as its delegate to seek the co-operation of the American Institute of Homœopathy. The invitation which, on my way, has diverted my steps to Boston has brought me many gratifications. But I shall be best rewarded for any trouble I have taken if I can secure for my proposals the advocacy of those who have listened to me here, — of students, and professors, and physicians in general. Believe me that you will thereby be helping forward a work which will give us quite a new aspect in the eyes of the profession at large; which will make acceptance of homœopathy far readier, its study more easy, its practice more delightful and successful; and so will redound to the advantage of that common humanity for which each man should live, and in losing his individual life for which he most surely finds it.

INDEX.

---

www.ingramcontent.com/pod-product-compliance
Lightning Source LLC
Chambersburg PA
CBHW031408270326
41929CB00010BA/1380